Handbook of Family Pla

CW00369257

With
Compliments

Wyeth

W O R L D   L E A D E R S
I N   O R A L   C O N T R A C E P T I O N

Wyeth Laboratories, John Wyeth & Brother Ltd.,
Taplow, Maidenhead, Berks SL6 0PH.

Handbook of Family Planning

Edited by

Nancy Loudon MB, ChB

Medical Co-ordinator, Family Planning and Well Woman Services Lothian Health Board; Lecturer, Department of Obstetrics and Gynaecology, University of Edinburgh

IN COLLABORATION WITH

John Newton MD, FRCOG
Professor of Obstetrics and Gynaecology, University of Birmingham

FOREWORD BY

David T. Baird DSc, FRCP, FRCOG
Professor of Obstetrics and Gynaecology, University of Edinburgh

CHURCHILL LIVINGSTONE
EDINBURGH LONDON MELBOURNE AND NEW YORK 1985

CHURCHILL LIVINGSTONE
Medical Division of Longman Group Limited

Distributed in the United States of America by
Churchill Livingstone Inc., 1560 Broadway, New York,
N.Y. 10036, and by associated companies, branches and
representatives throughout the world.

First published 1985

ISBN 0 443 02480 4

British Library Cataloguing in Publication Data

Handbook of family planning.
  1. Contraception
  I. Loudon, Nancy B, II. Newton, J. R.
  613.9′4   RG136

Printed in Great Britain by
Butler & Tanner Ltd, Frome and London

Foreword

The inclusion of Family Planning within the general medical services is a relatively new development in the UK, and still an exception in most countries of the world. The reasons for this are many, and include religious objections, social institutions and political fear that increasing access to contraception would lead to a decline in population and widespread promiscuity. With a few notable exceptions the medical profession was very reluctant to accept responsibility for this area of 'social' medicine and, prior to 1950, very few medical schools in the UK taught family planning to undergraduates.

The change which has occurred in the UK in the last 15 years has been dramatic, such that every practising doctor is expected to have some working knowledge of contraceptive practice. Undergraduate courses of obstetrics and gynaecology include family planning as an important objective, and expect every graduate to have some knowledge in this area. It is now widely accepted that only by easy access to all aspects of family planning can individual men and women have the ability to control their reproductive life. Ten years ago this was recognised at national level with the passing of the National Health Service Reorganisation Act 1973, which made provision for free family planning advice within the National Health Service.

At the same time, partially in response to the pressures of an ever expanding population of the world, but also because the present methods are not entirely adequate, there has been increasing research and development into better contraceptive methods. Nancy Loudon and her team of authors have provided a practical up-to-date account of family planning methods and services, which is aimed at providing a handbook for doctors and nurses in training and thereafter in practice. The training requirements for those in different branches of medicine are spelt out in detail as well as the theoretical and practical advantages and disadvantages of the various methods. It is encouraging to recognise that Britain continues to produce workers of such high quality, to follow in the footsteps of such pioneers as Marie Stopes.

Edinburgh   David T. Baird
1985

Preface

In the United Kingdom last year 2.7 million women sought contraceptive advice from general practitioners and 1.7 million from family planning clinics. Still others did so in hospital, at gynaecological and at postnatal clinics, and I hope during the course of consultation with doctors in many other specialties. Now men too are taking responsibility for contraception and often come with their partners to seek advice.

When the Family Planning Association took the initiative and introduced theoretical and practical training for doctors and nurses in the 1950s, its Medical Department produced a Clinic Handbook to complement their training programmes. The book became an invaluable asset to the staff of family planning clinics throughout the country and although some sections were updated from time to time only one edition was published. Replacement became urgent and the National Association of Family Planning Doctors was asked to re-write it. In response to that request this Handbook of Family Planning has been written by members of NAFPD and invited authors.

I am indebted to Professor John Newton for the help and advice he gave me in planning the book and with the early editing of some chapters. Since the handbook is intended primarily for use by family planning doctors, general practitioners, gynaecologists in training and nurses I felt it important to try and ensure that the needs of these different groups were catered for as far as possible. I was very fortunate that Mr James Drife, Mrs Margaret Foxwell, Dr Alistair Moulds, Mrs Mary Rankin and Dr Elizabeth Wilson undertook the task of reading the manuscript with their own specialties particularly in mind. I greatly appreciated the many helpful suggestions they made which improved the format of the book and increased its relevance to their colleagues.

To Mrs Sandra McDonagh who typed the manuscript so meticulously I am most grateful, especially for the stoical way in which she re-typed large sections of many chapters to incorporate new material as it appeared.

The Council of NAFPD paid me the compliment of asking me to edit this handbook and although during the past 2 years it seemed to occupy all my free time and sorely tried the patience of my husband and family I could not have undertaken the task without their help. My husband, John, whose advice and encouragement made it all possible, has at the same time been my strictest critic correcting and arguing about many points which otherwise might have gone unchallenged. To him I owe more thanks than to anyone else particularly for his understanding when household chores and social commitments had to take second place to 'the book'. I also appreciate the encouragement and support which my sons, Alasdair and Richard, and my daughter-in-law, Mary, have given me and for their helpful advice particularly on points of Scottish Law.

To Anne Barrett, who drew the illustrations so carefully, and to Sylvia Hull and Mary Lindsay of Churchill Livingstone, who guided me very patiently through the pitfalls of editing my first book, I owe a great debt of gratitude.

In a scene which changes as rapidly as contraceptive practice the editing of this book has been a constant battle to keep abreast of recent advances. New preparations and devices have been developed, old ones abandoned. Opinions and practices have changed, ethical and legal concepts have been questioned and new papers, which just had to be quoted as references have constantly appeared in the literature. Only the co-operation of the authors and their willingness to re-write many sections of their chapters allowed me to cope with these developments and produce an up-to-date handbook. To all of them I would like to express my sincerest thanks and appreciation. I know they will all feel it has been worthwhile if this handbook proves a quick helpful reference manual for busy doctors and nurses to enable them to provide their patients with sound advice in choosing and using happily a safe and effective method of contraception.

Edinburgh   Nancy B. Loudon
1985

Contributors

Dame Josephine Barnes DBE, MA, DM, FRCP, FRCS, FRCOG
Consulting Obstetrician and Gynaecologist, Charing Cross Hospital and Elizabeth Garrett Anderson Hospital; President, National Association of Family Planning Doctors

Julian A. Elias FRCOG
Consultant Obstetrician and Gynaecologist, Greenwich Health District; Honorary Senior Lecturer, King's College Hospital, London

Max Elstein MD, FRCOG
Professor of Obstetrics and Gynaecology, University of Manchester

G. Marcus Filshie FRCOG
Senior Lecturer and Honorary Consultant, Department of Obstetrics and Gynaecology, University of Nottingham

Anna M. Flynn DPH, MRCOG
Senior Clinical Research Fellow in Natural Family Planning, Department of Obstetrics and Gynaecology, University of Birmingham

Judy Greenwood MB, MRCPsych
Fellow in Community Psychiatry and Psychosexual Counsellor, Royal Edinburgh Hospital

Elizabeth H. Gregson MB, ChB
Lecturer in Family Planning, Department of Obstetrics and Gynaecology, University of Bristol; Vice Chairman, National Association of Family Planning Doctors

John Guillebaud MA, FRCS, FRCOG
Senior Lecturer, Academic Unit of Obstetrics and Gynaecology, Middlesex Hospital; Medical Director, Margaret Pyke Centre, London

Patricia A. Last FRCS, FRCOG
Lecturer, Department of Gynaecology, St Bartholomew's Hospital; Head, Women's Unit, BUPA Medical Centres

John D. O. Loudon FRCS, FRCOG
Consultant Obstetrician and Gynaecologist, Eastern General Hospital, Edinburgh; Senior Lecturer, Department of Obstetrics and Gynaecology, University of Edinburgh

Nancy B. Loudon MB, ChB

Medical Co-ordinator, Family Planning and Well Woman Services Lothian Health Board; Lecturer, Department of Obstetrics and Gynaecology, University of Edinburgh

Malcolm C. Macnaughton MD, FRCP, PRCOG, FRSE

Muirhead Professor of Obstetrics and Gynaecology, University of Glasgow

The late George M. Morris MB, BS, LRCP, MRCS, D(Obst)RCOG, MRCGP

Formerly General Practitioner; Senior Lecturer GP Unit, University College Hospital; Clinical Assistant, Hornsey Central Hospital, London

Alistair J. Moulds MB, ChB, MRCGP

General Practitioner, Basildon, Essex

John R. Newton MD, FRCOG

Professor of Obstetrics and Gynaecology, University of Birmingham

Kay M. Reid MRCS, LRCP

Senior Clinical Medical Officer (Family Planning), North Derbyshire Health Authority

Robert Snowden BA, PhD, DSA

Director, Institute of Population Studies, University of Exeter

Elizabeth S. B. Wilson MB, BS

Area Co-ordinator Family Planning Services, Greater Glasgow Health Board

Contents

1   *John D. O. Loudon*

Family planning in the United Kingdom: services and training

HISTORICAL BACKGROUND

Family planning services

In 1946 the National Health Service Act empowered local health authorities and regional boards to open contraceptive clinics where advice could be provided for nursing mothers seeking help with family planning on medical grounds, or to contribute to organisations which undertook to give such advice. Few health authorities took any action and voluntary organisations, particularly the Family Planning Association (FPA), were left to develop contraceptive services by opening clinics throughout the country. These clinics were used almost exclusively by married women, the majority of whom had to pay for the service and for contraceptive supplies, although payment for both was often waived for those in need.

Family limitation was achieved by abstinence, the rhythm method, coitus interruptus, chemical contraceptives, occlusive diaphragms, condoms, or by illegal abortion. Therapeutic termination of pregnancy

1

was rare. Sterilisation was unusual, the majority of operations being performed by tubal ligation in the puerperium or at caesarean section.

In the 1960s, with the introduction of oral contraception, modern intrauterine devices and the perfection of laparoscopy and vasectomy techniques permitting sterilisation on a large scale, the scene gradually changed. Contraception was openly discussed and these 'respectable' medical methods found much more favour with doctors. Changing attitudes to premarital chastity led to a demand by the unmarried for effective contraception.

Abortion Act (1967). Implemented April 1968

This acted as a real spur, for it was recognised that unless good contraception was freely available the demand for termination of unwanted pregnancies would escalate (Ch. 12).

National Health Service (Family Planning) Act (1967)

This enabled local authorities in England and Wales to give contraceptive advice, supplies and appliances to persons seeking contraception, from whom charges could be recovered. A similar service was introduced into Scotland in 1968 by the Public Health and Health Services Act.

These acts were liberally interpreted by local authorities and, as a result of agency arrangements with the FPA, free contraception was provided for many women.

National Health Service Reorganisation Act (1973)

A completely free family planning service at clinics and hospitals came with the reorganisation of the National Health Service in 1974. This was extended into primary care in July 1975.

Thus, over a decade Family Planning Services were revolutionised so that in the United Kingdom there exists the first completely free, universally available contraceptive service in Western Europe.

PRESENT POSITION

1. Free contraceptive services are available from general practitioners, in family planning clinics, in hospitals and from some voluntary organisations such as the FPA and Brook Advisory Centres which receive grants from Central Government or Health Authorities.

2. Patients have freedom of choice as to where they go for contraceptive advice. They can attend their own general practitioner, another general practitioner if their own declines to provide the service, or a family planning clinic, to which they have open access and do not require referral.

3. General practitioners on the contraceptive list are identified with a 'C' after their names on the list of family doctors, which is available to the public in post offices, public libraries, etc.

4. Male and female sterilisation is available free in NHS hospitals. Vasectomies are also carried out in some of the larger family planning clinics, and by a small number of general practitioners in the surgery.

5. No charge is made for termination of pregnancy under the NHS, although facilities vary widely in different parts of the country. In some areas patients have to resort to seeking abortions from gynaecologists in private practice or from voluntary agencies, such as the British Pregnancy Advisory Service (BPAS), which also undertakes abortions on an agency basis for at least one health authority.

6. All contraceptive supplies are available free of charge at family planning clinics.

7. All contraceptive supplies except condoms may be prescribed by general practitioners under the National Health Service and are exempt from prescription charge.

8. Item of service payments are made to general practitioners and to hospital staff (Ch. 17).

Other services

In addition to contraception many family planning clinics and general practitioners now provide other help, advice and services in this field.

1. Counselling for those with personal, marital or sexual problems (Ch. 16).

2. Health screening during the family planning consultation (Ch. 15).

3. Special advisory services for young people (Ch. 2).

4. Advice, counselling and ongoing support for women before and after termination of pregnancy (Ch. 12).

5. A domiciliary service (p. 4).

6. Pregnancy testing.

7. Postcoital contraception (Ch. 11).

8. Health education (Ch. 15).

9. Larger clinics act as resource centres for family planning literature, statistics etc.

10. Training facilities for medical and nursing staff and members of other disciplines.

11. Programmes of clinical and epidemiological research.

Family Planning Association (FPA)

In 1930 the National Birth Control Council was formed to co-ordinate the work of various birth control societies. In 1939 this became the Family Planning Association. The FPA was thereafter the major source of provision of contraceptive advice in this country. This continued effectively until 1974 when the National Health Service was reorganised and provided a free comprehensive family planning service. The FPA, having thus attained its primary objective, decided that its future role would lie in the fields of public information and education. Thus it remains in being, once more largely a voluntary organisation, providing a very useful service.

Brook Advisory Centres

The first Brook Advisory Clinic was opened in London in 1964 with the aim of 'the prevention and mitigation of the suffering caused by unwanted pregnancy by educating young persons in matters of sex and contraception and developing among them a sense of responsibility in regard to sexual behaviour.' Since then centres have been set up in Birmingham, Avon, Coventry, Edinburgh and Merseyside. At these clinics young people find sympathetic advice and services to help them with their contraceptive problems and their sexual and emotional relationships.

Domiciliary family planning

The majority of couples are sufficiently motivated to attend their general practitioner or family planning clinic to obtain contraceptive advice. However, for the small percentage of the population who lack this motivation or are unable for some other reason to obtain help, domiciliary family planning services serve a useful function. Doctors and nurses visit patients in their homes or in some areas accompany them to a clinic.

National Association of Family Planning Doctors (NAFPD)

In 1974 the service role of the FPA was largely taken over by the NHS. Until then doctors and nurses working within family planning services had been members of medical and nursing national bodies of the FPA. The need for organisations to replace these bodies became obvious and the National Association of Family Planning Doctors was founded. Its objects are:

1. To advance the education of medical practitioners in all matters appertaining to family planning and sexual medicine.

2. To facilitate the study of these matters and the exchange of information.

3. To promote research in various ways.

4. To promote high standards of training and practice.

NAFPD is now a flourishing body to which doctors working in family planning clinics together with many general practitioners, gynaecologists and others interested in family planning belong. Its aims are fulfilled by the organisation of symposia and by publishing a journal, the *British Journal of Family Planning* *(BJFP)* containing authoritative and up-to-date articles and comments on the subject. By its membership of the Joint Committee on Contraception (see below), it has considerable influence on standards of training and assessment.

National Association of Family Planning Nurses (NAFPN)

NAFPN and the Scottish Society of Family Planning Nurses perform similar functions for nurses. Members are entitled to copies of *BJFP* at a reduced rate.

TRAINING IN FAMILY PLANNING

Doctors

Fertility control now features prominently in the undergraduate teaching programme in most medical schools, and doctors entering practice should have a good basic knowledge of contraception.

Joint Committee on Contraception (JCC)

In 1972 when family planning was increasingly provided by local health authorities, the Royal College of Obstetricians and Gynaecologists (RCOG) took the initiative of setting up a committee with the Royal College of General Practitioners, (RCGP) the FPA, the

Faculty of Community Medicine and later NAFPD in order to supervise standards of training in family planning and certification of doctors. The Joint Committee on Contraception has now assumed sole responsibility in this field.

Basic requirements. Those wishing to acquire the certificate of the Joint Committee on Contraception should have full registration with the General Medical Council (GMC) (though special arrangements can be made for training those with only limited registration) and must be under the age of 65. They are expected to have acquired the basic skills of gynaecological examination, preferably by working in a hospital post approved by the RCOG for its diplomas, although provision is made for this requirement to be met in other ways (Appendix 1).

Theoretical training. The candidate must attend a course of lectures designed to meet the requirements of JCC. These courses are organised by directors of family planning services, postgraduate deans and tutors, and others, who are expected to submit the details of their planned courses to JCC for approval and subsequently to send a list of those who attend the course (Appendix 2).

Guidelines for basic courses are available free of charge from the Secretary of JCC or from NAFPD.

Practical training. Within two years of attending the theoretical course the trainee should undergo practical instruction at a recognised clinic under the guidance of an instructing doctor, who must be recognised as such by JCC. In the past most of the recognised places for training have been family planning clinics in the community and in hospitals, but increasingly health centres and general practitioners working within them have received recognition (Appendix 3). Unless the trainee has had prior experience in practical family planning, a minimum of eight sessions of training will be required. This training includes acquisition of the ability to insert intrauterine devices. Sufficient experience in this is difficult to obtain in some clinics and JCC therefore recognises groups of clinics which can provide complementary experience.

Practical training experience is recorded on JCC form L by the instructing doctor. This is a confidential document and is sent to JCC direct (Appendix 4).

Instructing doctors. Details of training, application for recognition and advice to instructing doctors are contained in Appendices 5, 6 and 7.

The instructing doctor is responsible for:

1. Checking, before practical training starts, that each trainee is

registered with the GMC and has the necessary basic skill in gynaecological examination.

2. Providing training in all types of contraceptive techniques and counselling, including the insertion of intrauterine devices.

3. Carrying out continuing assessment of the trainee's abilities and filling in form L after every session.

4. Assessing, during the planned final session, the trainee's competence and, if satisfied, completing the certificate on form L and informing the candidate that he/she may apply to JCC for a certificate.

5. Advising a trainee, who has not attained sufficient competence, that further practical training is necessary or, alternatively, requesting independent assessment by a senior colleague.

Reaccreditation. Instructing doctors have to apply for 'continuing recognition' every five years, using form F2, providing evidence that they have attended an updating course. (Appendix 8).

Obtaining the certificate. Trainees who have been informed that they have acquired the necessary skill should complete JCC form J and send it with the current fee to the secretary of the Joint Committee on Contraception (Appendix 9).

At first only a minority of established general practitioners obtained certification, although a great many more attended the theoretical course without completing practical instruction. The theoretical course is gradually becoming accepted as part of vocational training for general practice and a publication from the Royal College of General Practitioners (1981) advocated that all trainees should in future obtain the JCC certificate. Although general practitioners are not required to hold a certificate to provide contraceptive advice, most health authorities expect doctors working in their clinics to do so.

Regulations for both the Diploma and the Membership of the RCOG require the doctor to undertake instruction at not fewer than eight practical training sessions in family planning, but not necessarily the theoretical course.

Nurses and midwives

In 1967 the General Nursing Council agreed to include family planning in its training syllabus and there can now be few, if any, courses of training for student nurses or midwives in the United Kingdom which do not include some instruction in contraception.

Postbasic courses (Course 900), formerly the responsibility of the Joint Board of Clinical Nursing Studies in England and Wales and the

Committee for Clinical Nursing Studies in Scotland, are now organised by the National Boards of each country. They are designed a) to prepare nurses and midwives to work in family planning clinics, either specialist family planning clinics or those in general practice or in hospital; b) to give instruction in family planning to nursing and midwifery tutors and to nurses working in departments where patients are particularly likely to benefit from family planning advice, e.g. gynaecological and postnatal wards. Health visitors, whose work is principally in the community, also attend these courses in some areas.

Details of Course 900 are contained in Appendix 10.

Training in psychosexual medicine

Institute of Psychosexual Medicine

This is an organisation for promotion of psychosexual medicine through seminar training and research. It offers training to doctors to improve their skills with patients who seek help with sexual difficulties and related marital or psychosomatic problems. Membership is limited to medical practitioners. A certificate of competence is issued after completion of recognised courses of instruction.

Association of Sexual and Marital Therapists

The aims of this association are to promote the enrichment of people's sexual lives and to foster positive attitudes towards sexuality. Regular conferences and therapy workshops are held. Membership is open to those members of the helping professions who satisfy the training requirements of the Association.

Association to Aid the Sexual and Personal Relationships of the Disabled

Although this body is primarily concerned with supplying information for disabled people with sexual problems, it also helps to set up and programme study days for interested groups.

Training programmes are also organised by groups in different parts of the country, such as the Edinburgh Human Sexuality Group, which runs a 10-month (part-time) training course for doctors, nurses, psychologists, social workers and marriage guidance counsellors. Details of such groups can normally be obtained from local family planning services. A directory published by the *British Journal of*

Sexual Medicine lists FP clinics throughout the United Kingdom (*Referral Directory* 1982).

Addresses of organisations, associations, etc. mentioned in this chapter are to be found on pages 324–326.

REFERENCE

Royal College of General Practitioners 1981 Family planning. An exercise in preventive medicine. RCGP, London

Appendix 1

JOINT COMMITTEE ON CONTRACEPTION
27 Sussex Place, Regent's Park, London NW1 4RG — Tel. 01-262 5425

TRAINING REQUIREMENTS FOR DOCTORS WISHING TO OBTAIN THE J.C.C. CERTIFICATE

Both theoretical and practical training need to be undertaken during which assessment is carried out. The theoretical training should precede the practical which should follow within TWO years.

ELIGIBILITY
FULL REGISTRATION with the GENERAL MEDICAL COUNCIL U.K. is a basic requirement, AND APPLICANTS SHOULD BE UNDER THE AGE OF 65.
Those who have only limited registration may attend theoretical courses, but if they wish to proceed to the practical training and subsequent certification they must submit full details of their proposed programme well in advance to the G.M.C. where their cases will be considered on individual merit.
Before training it is necessary to obtain Form J from the JCC, which has the following three sections, A.B.C. Of these Part A must be completed before practical training is initiated.

PART A. BASIC GYNAECOLOGICAL EXAMINATION
Some doctors have not had sufficient experience to acquire the necessary skills to carry out the basic gynaecological examination which is essential to all contraceptive practice. It is important that trainees in contraceptive practice should acquire this experience before embarking on practical training. There are THREE alternative possible ways of certifying this:—
- (i) By holding or having held an obstetric & gynaecological post, which is recognised for training for the DRCOG or for Membership of the RCOG or RCGP.
- or (ii) By providing evidence of attendance at a minimum of FOUR sessions at a gynaecological clinic under the direction of a consultant gynaecologist or recognized JCC family planning instructing doctor at which the following techniques have been practised:—
  (a) Digital and speculum examination of the vagina.
  (b) Vaginal and cervical cytology procedures.
  (c) Pelvic examination.
  (d) Breast examination.
  The gynaecologist will be asked to certify that a satisfactory standard has been reached.
  Trainees need to arrange this in advance by personal arrangement. This is not always easy to obtain.
- or (iii) By self certification in which the trainee should give details of previous gynaecological experience. Trainees are warned that there is no available time in family planning clinics to teach basic skills and if a trainee is found to have insufficient expertise, further training will need to be delayed until section ii (above) has been completed.

PART B. THEORETICAL INSTRUCTION
This involves attendance at a training course recognised by the J.C.C., and its number must be recorded on the application form. Courses are of a minimum of eight hours during which the following subjects must be studied:—
- (i) Acceptability and motivation in contraception.
- (ii) "Choice of Method" and discussion of "Risks and Benefits."

(iii)   All available contraceptive methods. Advantages/disadvantages, and risk/benefit ratios, reliability. Necessary technical skills and management of associated complications including resuscitation.
(iv)   Sterilising procedures.
(v)   Abortion counselling, legal aspects, techniques.
(vi)   Emotional and psychological factors related to fertility regulation.
(vii) Sexually transmitted diseases.
(viii) Family Planning services — organisation and administration — community, general practice and hospital.
Flexibility in organisation of such courses in accordance with local needs is encouraged.

PART C. PRACTICAL INSTRUCTION

Experience of working within a family planning training clinic under the supervision of an instructing doctor who is recognised by the J.C.C. is necessary. The place where this is undertaken should be recorded on the application form. Full understanding of the prescription of all methods of birth control including ability to insert IUCDs and fit vaginal barriers is necessary. These as well as the ability to instruct patients will be assessed by the instructing doctor. A confidential report will be submitted to the J.C.C. upon receipt of which the eligibility for a certificate will be decided.

While there is no specified maximum or minimum number of practical sessions, most trainees will need to attend at least EIGHT. Fewer than four attendances would be accepted only in exceptional circumstances.

While previous attendance as an observer at family planning clinics prior to the DRCOG examination will be taken into consideration, it does not exempt the trainee from practical instruction.

At the conclusion of training, and subject to a satisfactory assessment, a completed application form should be sent to the Secretary of the J.C.C. together with the appropriate fee and a certificate will then be issued.

Appendix 2

JOINT COMMITTEE ON CONTRACEPTION
27 Sussex Place, Regent's Park, London NW1 4RG — Tel. 01-262 5425   **D**

APPLICATION FOR RECOGNITION OF A COURSE IN THEORETICAL INSTRUCTION FOR CERTIFICATE IN CONTRACEPTION AND FAMILY PLANNING

It is my intention to arrange and conduct a course of theoretical instruction for the above purpose. A copy of the syllabus with names of the lecturers is attached. I hereby apply for formal recognition of this course for the purpose of training for the Certificate in Contraception and Family Planning.

When accepting trainees for the course I shall provide each with copies of the Training Syllabus for Doctors and the Application Form. Please send me .... copies of each for this purpose.

At the conclusion of the course I will submit a list of the names of those doctors who attend the full course and will notify each trainee of the official number of the course.

SIGNED ....

NAME ....

(printed please) DESIGNATION ....

ADDRESS ....

            ....

            ....

TELEPHONE NUMBER ....

DATE OF COURSE ....

AUTHORITY ON WHOSE BEHALF COURSE IS BEING ORGANIZED ....

ADDRESS ....

            ....

            ....

A new application for recognition should be made for each separate course of theoretical instruction

Please send this application together with the copy of the syllabus to:

The Secretary,
Joint Committee on Contraception,
27 Sussex Place,
Regent's Park,
London NW1 4RG

FOR OFFICE USE
No. allocated to
course ....

Appendix 3

JOINT COMMITTEE ON CONTRACEPTION
27 Sussex Place, Regent's Park, London NW1 4RG — Tel. 01-262 5425

PRACTICAL TRAINING FOR THE J.C.C. CERTIFICATE
WITHIN GENERAL PRACTICE

1. To ensure the maintenance of uniform high standards and to conform to existing J.C.C. requirements for clinics, the following criteria would need to be fulfilled:
    (i)   General Practitioners who hold a J.C.C. Certificate and have been engaged in continuous family planning practice for at least three years may proceed to recognition as an Instructing Doctor after attending an appropriate approved Instructing Doctors' course and assessment.
    (ii) The usual confidential reports on the trainee will be completed.
    (iii) General Practice premises should meet the required standards and may be visited before approval.*
2. Most of the practical training should be undertaken at approved family planning training sessions, provided a wide variety of experience is obtainable in a relatively short period.
3. While a minimum of eight sessions is usually required, a J.C.C. trainee who acquires a great deal of family planning experience within general practice may require fewer clinic sessions for training.
4. The final assessment of trainees for the J.C.C. Certificate will take place at an approved family planning training session.
5. Application for practical training for the J.C.C. Certificate would involve all the existing preliminaries as laid down in the J.C.C. regulations.

---

*   Details of and application forms for recognition of clinic and practice premises for training purposes are available from the Secretary to JCC at the above address.

Appendix 4

JOINT COMMITTEE ON CONTRACEPTION
27 Sussex Place, Regent's Park, London NW1 4RG — Tel. 01-262 5425

L

CONFIDENTIAL REPORT ON TRAINEE FOR CERTIFICATE IN CONTRACEPTION AND FAMILY PLANNING

FULL NAME OF TRAINEE........................DATE OF BIRTH..............

ADDRESS ....

....

FULL ADDRESS OF CLINIC..

No. of Sess.	Date	COMMENTS—including methods dealt with at particular session and on the ability and approach of the trainee.	No. of IUCD Insertions	Instructing Doctor's Name JCC No.	Signature
1.					
2.					
3.					
4.					
5.					
6.					
7.					
8.					

I certify that in my opinion the forenamed candidate:
*(a) has achieved competence in the techniques of family planning and I recommend that he/she be granted the certificate of the Joint Committee on Contraception.
or *(b) has not reached the standard required for the certificate of the Joint Committee on Contraception.
or *(c) has been advised to undertake more training sessions.
*delete as appropriate

Date...........................Signed ....

Certifying doctor's name and number ....
FOOTNOTE: Details of training sessions carried out at clinics other than the clinic at which this report is initiated should be filled in on separate forms and returned to the original clinic. It is accepted that eight sessions may not be adequate to achieve full competence in all contraceptive techniques. Additional forms should be used if necessary.

On completion, this and any supplementary forms should be forwarded to:
The Secretary
Joint Committee on Contraception
27 Sussex Place
Regent's Park
London NW1 4RG

Appendix 5

JOINT COMMITTEE ON CONTRACEPTION
27 Sussex Place, Regent's Park, London NW1 4RG — Tel. 01-262 5425

TRAINING OF DOCTORS WISHING TO BECOME INSTRUCTORS IN FAMILY PLANNING
GUIDELINES FOR EMPLOYING AUTHORITIES

The responsibility for selection of suitable candidates to train as Instructors should normally be the responsibility of their employing authority. Such candidates may include clinic medical officers and family practitioners working in family planning clinics or their own premises.
   Doctors applying to become Instructors should:
   (i)   Have a JCC certificate or its equivalent
   (ii)   Have been in continuous family planning practice for at least three years
   (iii)   Have an interest in teaching.
   The requirements for recognition as an Instructing Doctor are attendance at a course recognised by the JCC, and assessment after completion of the course. Doctors holding the MRCOG, with practical experience in family planning work, should attend an approved course but may then be accepted as Instructing Doctors on direct application to the JCC, without formal assessment. Specialists holding an MRCOG degree and currently engaged in teaching family planning practice to doctors in hospitals will not be required to attend an instructing doctors' course or undergo assessment, and will normally be recognised as Instructing Doctors by the JCC.

Courses for Applicants for Recognition as Instructing Doctors
The responsibiity for providing and financing appropriate training will normally be that of the Area or Regional Health Authority, in conjunction with local postgraduate training organisers. Guidelines, together with advice and support in the provision of training courses, may be obtained from the National Association of Family Planning Doctors. The programme must be approved in advance by a special panel of the JCC. Recognition of individual courses may be obtained by submission to the JCC, but submission of annual training programmes is preferable.
   Courses must be of a minimum length of 5 half-day sessions, but these need not be consecutive. The content of the course must be concerned with (a) updating (b) teaching techniques, with emphasis on one to one and small group teaching.
   The following list of headings indicates the scope required in the teaching of Instructing Doctors:
   (a)   Clinical practice, including working with a family planning nurse
   (b)   Up-to-date knowledge of all accepted contraceptive techniques including procedures for sterilisation; acquaintance with research projects in progress locally; knowledge of aspects of family planning procedures in which it is hoped scientific developments will occur.
   (c)   Special aspects of patient care including:
              —the young teenager
              —the menopausal patient
              —psychosexual problems
              —sexually transmitted diseases
              —screening for pelvic cancer
              —cultural and religious influences.
   (d)   Interviewing, consulation, counselling and teaching techniques.
   Organisers of courses for Instructing Doctors will be required to submit to the JCC the names of doctors who have attended satisfactorily.

Assessment

Organisers of courses should arrange for assessment of potential Instructing Doctors on completion of the course. The JCC may be able to assist Health Authorities to find suitable assessors from neighbouring areas, but normally they will be drawn from the senior family planning doctors of the Area or Region. Assessment is required to be conducted outside the theoretical course and will be in two parts:

(i)   In the centre in which the applicant can conduct clinical training of doctors

(ii)   By interview.

Assessors should refer to the list of topics (a) to (d) above. The names of successful candidates should be reported to the JCC. Candidates who fail may be offered further training opportunities and subsequent assessment may be carried out by a different assessor.

Appendix 6

JOINT COMMITTEE ON CONTRACEPTION   **F1**
27 Sussex Place, Regent's Park, London NW1 4RG — Tel. 01-262 5425

APPLICATION FOR FIRST RECOGNITION AS INSTRUCTING DOCTOR FOR CERTIFICATE IN CONTRACEPTION AND FAMILY PLANNING

NAME OF APPLICANT..

QUALIFICATIONS...

ADDRESS ....

....

....

ADDRESS OF CLINIC..

....

CERTIFICATE OF FAMILY PLANNING ISSUED BY..............................

NUMBER OF CERTIFICATE...........DATED..................................

IF SEPARATE I.U.C.D. CERTIFICATE, NO. ....DATED.............

IS THE APPLICANT A RECOGNIZED FPA INSTRUCTING DOCTOR?   YES/NO

IF YES, NUMBER OF INSTRUCTING DOCTOR'S CERTIFICATE....................

DATED......................

DATE OF INSTRUCTING DOCTOR COURSE.....................................

JCC REGISTRATION NO. ....

PLACE OF COURSE...

I assessed the above named doctor on......................at ....

He/she has been in continuous family planning practice during the last.............years.

In my opinion he/she *is now competent to be an instructing doctor

            *requires further experience before becoming an instructing doctor

            *delete as appropriate

General comment:

Signed ....Name ....
                                     (typescript or block letters please)

Designation.................Address ....

Appendix 7

JOINT COMMITTEE ON CONTRACEPTION
27 Sussex Place, Regent's Park, London NW1 4RG — Tel. 01-262 5425

ADVICE TO INSTRUCTING DOCTORS

The sponsoring organizations are most anxious that their Certificate in Contraception and Family Planning is held in high regard and that the holders of the Certificate are competent in the full range of non-surgical contraceptive techniques.

The instructing doctor should also ensure that the trainee is **FULLY REGISTERED** with the General Medical Council (U.K.).

Those who have only limited registration may attend theoretical courses, but if they wish to proceed to the practical training and subsequent certification they must submit full details of their proposed programme well in advance to the G.M.C. where their cases will be considered on individual merit.

Before commencing practical training the trainee will have a knowledge of basic gynaecological examination and the instructing doctor should inspect the certificate confirming this. (A form of self-certification for experienced practitioners has been adopted. An instructing doctor must advise such a doctor if it is felt that his/her basic skill is insufficient to allow efficient practical training). He/she will also be undergoing or have completed a course of theoretical instruction covering contraception, family planning and related subjects. It is the task of the instructing doctor to provide instruction in all types of contraceptive techniques — choice of method (including sterilisation) related to indication and contra-indication; complications and their management; fitting of occlusive diaphragms and insertion of intra-uterine devices etc. in accordance with the current syllabus.

After the first session of practical instruction, the instructing doctor will initiate a confidential report on the trainee and will complete the appropriate section detailing the range of methods studied and the trainee's approach and ability. At the conclusion of each subsequent training session the instructor will complete the appropriate section with similar information. If a trainee carries out any instruction sessions at a clinic other than the one at which practical instruction was commenced, a separate form should be completed for that session and returned to the original clinic for attachment to the principal form. The Committee have not laid down a minimum number of practical sessions for each trainee, but fewer than four would be accepted only in exceptional circumstances, for instance if the trainee has previous experience. Eight would be a reasonable number to expect for the complete novice. The Committee do not intend that there will be any formal examination so that it will be the responsibility of the instructing doctor to assess each trainee's competence. Only when absolutely satisfied of that competence and after Part B has been completed should the instructor complete the confidential report and forward it to the secretary of the Joint Committee, at the same time informing the trainee that he/she may now make formal application to the Committee for a Certificate.

At the conclusion of the intended final session an instructing doctor may feel that a trainee has not attained sufficient competence to permit recommendation for a Certificate. In that event the instructing doctor may suggest further training sessions or, alternatively, may request independent assessment by a senior colleague. The Specialist in Community Medicine in charge of family planning may be able to help to find such a colleague. If no appropriate senior instructing doctor is available the instructor may seek the advice of the Joint Committee who will arrange appropriate assessment.

Appendix 8

JOINT COMMITTEE ON CONTRACEPTION
27 Sussex Place, Regent's Park, London NW1 4RG — Tel. 01-262 5425

F2

APPLICATION FOR CONTINUING RECOGNITION AS AN INSTRUCTING DOCTOR FOR THE CERTIFICATE IN CONTRACEPTION AND FAMILY PLANNING

(To be completed in typescript or block letters)

Name.................................Date of birth ....

Qualifications..

Address...

....

Address of training centre(s) where instructing ....

....

Certificate of family planning issued by ....

Number of certificate...........Date of issue..

If separate IUCD Certificate No...........Date of issue ....

Date of previous "continuing recognition" (If applicable) ....

Instructing doctor No..............Date of issue......................................

Date and place of updating course ....

Summary of content of updating course ....

....

....

I...certify that I have been in continuous

family planning practice for............years and have been instructing in family planning

for............years.

My employing authority is ....

....

PLEASE RETURN TOP COPY TO:
The Secretary
Joint Committee on Contraception
27 Sussex Place
Regent's Park
London, NW1 4RG

Signed...............................

Designation ....

Date ....

Appendix 9

JOINT COMMITTEE ON CONTRACEPTION
27 Sussex Place, Regent's Park, London NW1 4RG — Tel. 01-262 5425

J

APPLICATION FOR CERTIFICATE IN CONTRACEPTION AND FAMILY PLANNING

Surname ....

Other names ....

Qualifications ....

I am fully registered with the G.M.C.   Number ....

Address ....

....

....

Telephone number ....
I hereby apply for the Certificate in Contraception and Family Planning having completed the
required training syllabus as detailed below. I enclose my cheque for £10.00.

SIGNED ....DATE ....

PART A—BASIC GYNAECOLOGICAL EXAMINATION

To be completed by applicant

*I held from ....to ....

a post as .... in

....Hospital which
is recognized by the Colleges for gynaecological training under the charge

of ....

....

*I attended .... sessions at the

gynaecological clinic under the charge of ....

....

To be completed by a consultant gynaecologist or recognised JCC instructing doctor.

*This is to certify that the trainee undertook the instruction as stated and that he/she is
competent in basic gynaecological examination.

SIGNED ....DESIGNATION ....

To be completed only by an experienced medical practitioner

*I consider that by virtue of my experience I have sufficient skill in basic gynaecological examination to allow me to embark on practical training.

SIGNED ....

*Please delete or complete as appropriate.

PART B—THEORETICAL INSTRUCTION

I undertook a course in theoretical instruction at ....

....

Reference number and date of course ....

PART C—PRACTICAL INSTRUCTION IN CONTRACEPTIVE TECHNIQUES

(Please give full address of clinic)
I completed practical instruction at ....

....

clinic on ....

A CERTIFICATE WILL BE DESPATCHED AFTER THE SATISFACTORY CONFIDENTIAL REPORT (FORM L) FROM YOUR INSTRUCTING DOCTOR HAS BEEN RECEIVED.

Please send this application to:
The Secretary,
Joint Committee on Contraception,
27 Sussex Place,
Regent's Park,
London NW1 4RG

and make the cheque payable to:
JOINT COMMITTEE ON CONTRACEPTION

Appendix 10

COURSE IN FAMILY PLANNING NURSING (COURSE 900).

Aim of the Course.
To train a Registered Nurse or Midwife to give family planning advice, treatment and care in a variety of situations such as hospitals, clinics, health centres, general practice premises and the individuals' own homes.

Entry Requirements.
The course is for registered Midwives or Nurses whose names are entered on the general part of the Register of the General Nursing Councils for England and Wales, Scotland and Northern Ireland.

Length of Course.
The minimum number of sessions to be attended shall be the equivalent of nine sessions of three hours duration. These sessions can be spread over a period of time which shall not exceed a period of twelve weeks.

Teaching Time and Clinical Experience.
The contents of the curriculum are divided into three units of learning and are so arranged that competence in skills can be gained on a sessional basis.

COURSE UNIT 1

Objectives

After completing this unit the nurse will be:—

1. Able to take, select and record accurately relevant aspects of the individual's medical and social history.
2. Aware of the contribution of family planning to individual and family health and to the well-being of society.

Skills	Knowledge	Attitudes
	Psychological, sociological and demographic implications of providing family planning services.	
	Family planning as a preventive health measure.	
Establishing and maintaining a relationship with the individual.	Psychology of interpersonal behaviour; the influence of verbal and non-verbal communication.	Accepts the importance of understanding and adapting to the individual's sensitivity.
Maintaining confidentiality and privacy for the individual.	Medico-legal aspects of the work.	Appreciates the confidential nature of the work and understands the need to develop insight and tact in giving a professional service.
Interviewing.	Interview techniques.	Is aware of the importance of listening.
Recognising hidden problems.	Socio-cultural and psychosexual background to sexuality.	Shows increasing sensitivity to human need and a sympathetic understanding of the values people hold when different from her own.
Accurate recording of data and information.	Current recording systems. Use of records in follow-up care.	Recognises the importance of correct data.

COURSE UNIT 2

Objectives

After completing this unit the nurse will:

1. Be competent to carry out an examination of the individual and to take cytological specimens.
2. Be able to interpret the findings of the individual's medical and social history and the findings of the physical examination and refer the individual to the doctor or other agency when necessary.
3. Have an understanding of the relevant aspects of anatomy, physiology and pathology.

Skills	Knowledge	Attitudes
Examination including:	Related aspects of anatomy, physiology and pathology.	Is aware of the need to show gentleness towards the vulnerable individual and respect for personal dignity.
relevant observations for possible pregnancy		
assessment of vagina		
visualization of the cervix using a speculum	Cervical cytology.	
taking cytological specimens.		
Teaching self-palpation of the breast.	Method of self-palpation of the breast.	
Appraisal, interpretation and selection for referral or advice.	Failure of method or use including the management of unplanned pregnancy.	
Referring whilst maintaining the relationship.	Organisation and availability of other services including psychosexual counselling, male and female sterilisation, termination of pregnancy, sub-fertility and special clinics.	

COURSE UNIT 3

Objectives

After completing this unit the nurse will be competent to:—
1.   Inform, advise and counsel the individual concerning the choice of methods.
2.   Teach the application of the chosen method.
3.   Re-appraise the chosen method.

Skills	Knowledge	Attitudes
Adopting a non-directive approach when informing, advising and counselling in relation to chosen method.	Related aspects of individual human rights.	Recognise the importance of impartial advice.
	All methods of contraception including efficiency, side effects and contra-indications.	
	Physiological and psychological aspects of male and female sterilisation.	
	Physiological, psychological and pathological aspects of termination of pregnancy, sub-fertility and sexually transmittable diseases.	
Teaching the application of:		Is willing to give understanding, support and guidance to individuals experiencing difficulty in learning about and accepting the chosen method.
mechanically occlusive methods		
spermicidal agents		
hormonal contraception		
physiological methods.		
Re-appraisal of chosen method.	Routine after-care.	

Choice of method

General principles of management
Assessment of factors which influence
  contraceptive choice
    Effectiveness
    Availability
    Acceptability
    Motivation
    Influence of friends
    Influence of the media
    Characteristics of the couple
  Medical assessment
    History
    Examination
  Provision of information to the patient
  Continuing support and follow-up
	Minor side effects
	    Checklist
	Groups with special needs
	  Young people
	  Older women
	  Mentally retarded patients
	  Physically handicapped patients
	  After delivery
	  After termination of pregnancy
	  Patients of a different culture
	  Poorly motivated patients
	Factors influencing the doctor and nurse
	  Effectiveness
	  Mortality and morbidity

The main object of a family planning consultation is to help the patient to choose the best method of contraception and to ensure that its use is as trouble-free as possible. It is usually the woman who consults the doctor for family planning advice because most 'medical' methods of contraception are used by the female partner. The right method for any individual or couple is the one that suits them best at that particular stage in their lives. Although any of the nonpermanent methods may be medically suitable, patients tend to fall into four categories:

1. *Those stating a preference* — new patients frequently state exactly which method they wish to use. However, it is important to establish how much each knows about the method requested, whether it really is the one they want, and also to ensure that the individual is not being unduly influenced by someone else.

2. *Those stating exclusions* — some patients are not sure what they want but seem quite clear about what they do not want. Any particular reason for the objection should be identified because it may be based on misunderstanding and the woman may change her mind when acquainted with the facts.

3. *Those who do not know* — these patients require a lot of information and guidance before they can begin to come to a decision.

4. *Those who find it difficult to communicate* — some patients are not very articulate, while others are intimidated or antagonised by what seems to them to be an authoritarian and alarmingly clinical atmosphere. They need to be given time to settle down, made to feel at ease and encouraged to express their views, fears and doubts.

A wide range of background factors particular to each individual influences the choice in every case. Many require special help and understanding.

GENERAL PRINCIPLES OF MANAGEMENT

Four main areas have to be considered:
1. Assessment of the factors which influence contraceptive choice.
2. Medical assessment.
3. Provision of relevant and accurate information.
4. Continuing support and follow up.

Assessment of the factors which influence contraceptive choice

Effectiveness

Maximum effectiveness is not always a top priority. Some women prefer to choose a method with a small chance of failure, admitting that although they do not wish a pregnancy they would not mind too much if this were to happen. For others pregnancy would be a major disaster and they may say that they would seek abortion if their contraceptive failed. For them effectiveness is of prime importance.

Availability

If a method is difficult to obtain the patient may decide against it or quickly abandon it after starting. If he or she has to be referred elsewhere to obtain the contraceptive of choice, arrangements should be made for them, or at least they should be told exactly what to do themselves. A reliable contraceptive should be supplied in the meantime.

Acceptability

The method must be acceptable to her and she must be able to manage the regimen required. Equally important is the acceptability of the method to her partner. In some relationships the man may make the decisions about contraception and send the woman to implement them.

Motivation

Motivation to choose and use contraception depends on the patient's interpretation of the effectiveness, availability and acceptability of the methods. This will be based on her own experience and on the information and advice available to and assimilated by her. Her decision and her ability to use a method successfully will be influenced by her personality, maturity, intelligence, emotional stability and confidence.

Influence of friends

The contraceptive experience of relatives, friends and neighbours may be a potent influence on choice, e.g. 'My friend got pregnant with a coil so I would never have that!'

Influence of the media

Television, radio, books, newspapers and magazines all have an enormous influence on contraceptive use. The popularity of methods swings according to current publicity. Care should be taken not to comply immediately with what may be an ill-chosen request for change of method before clarifying, and possibly countering, the reasons behind it.

Characteristics of the couple

1. *Their ages, health and the number and health of their children* should be recorded as well as *obstetric and contraceptive history.*
2. *Marital status.* Premarital sex and cohabiting couples are part of society today. Whilst some of these couples require a high degree of contraceptive efficiency others do not mind risking pregnancy. Indeed some girls may hope to become pregnant in order to advance a wedding date. Others may ask for contraception without apparent need — they do not have a sexual partner at the time but are aware that they may be at risk of 'unintended' intercourse. They should have an effective contraceptive to tide them over this period of uncertainty.
3. *The stability of the relationship.* This exerts a powerful influence on the contraceptive a patient chooses and her motivation to use it.

Details about her partner and their relationship are not always revealed. Thus one should be prepared to accept an apparently irrational decision and beware of making precipitate judgements on incomplete information.

Frequent requests for a change of method and recurrent problems with every one advised should alert the doctor to the possibility of an underlying sexual problem.

4. *Social status and social relationships.* The attitude to contraception of a happily married, financially secure mother of two will be different to that of a woman of the same age whose two children are in care, who has a series of temporary partners and low-paid employment. Yet both may be recorded in the notes as 'age 30, para 2'.

Women who feel isolated in a community may harbour anxieties about pregnancy, childbirth and contraception and need extra counselling and support.

Professional women often find it difficult to be in a non-controlling patient role. It should not be assumed they are knowledgeable about contraception though many of them are.

5. *The family background.* Relationships within the family, lifestyle, religion and culture all produce attitudes to and patterns for living which influence contraceptive needs and choice. Courses of action which appear ill-advised to the doctor may be wholly appropriate in the context of particular background circumstances.

Patients caring for parents, grandparents or others may postpone or limit their own family because, for the time being, it may seem impracticable to have a baby. They may need help to clarify their feelings and their plans.

6. *Finance, career, occupation and prospects.* The importance of these factors depends on personality and ambitions. To some it seems vital to postpone pregnancy until qualifications, job status or financial goals are attained. They may need a highly effective contraceptive or on the other hand help to dare to decide to have a baby.

Most unemployed couples see postponement of a pregnancy as a high priority. Yet some young girls, bored and lacking purpose, may see pregnancy and motherhood as a fulfilling experience and this is reflected in their attitude towards contraception.

Medical assessment

History

A history based on the following headings, *as appropriate to the individual patient,* should be taken at the first visit. In general practice much of this information will already be on record. It should be checked and updated if a change of method is considered and when the patient appears after a long absence. However, it is important to remember that the majority of patients attending for family planning are fit and many of the aspects of their health which have to be assessed are not concerned with illness.

1. Personal history. Age and occupation of the patient and her partner. Marital status. Proposed family size. Smoking habit.

2. Obstetric history. Number of pregnancies and their outcome. Difficulty in conceiving. Problems during pregnancy, delivery or the puerperium and whether the woman is breast feeding. Birth weight of children and their current state of health. Number of children in the family unit, i.e. the number of children the woman looks after.

3. Contraceptive history. Current method of contraception if any. Previous methods and the length of time they were used. Details of side effects and the reason for discontinuation. Unplanned pregnancies resulting from method failure.

4. Menstrual history. Date of last menstrual period and details of the menstrual cycle — its regularity, the amount and duration of blood loss and whether there is intermenstrual bleeding, dysmenorrhoea, or premenstrual tension. The previous menstrual pattern of those currently taking the pill or using an IUD. Age of menarche of young patients.

5. Gynaecological history. Dyspareunia. Cystitis, vaginal infections, and pelvic inflammatory disease with details of treatment. Gynaecological investigations, operations and their outcome.

6. General history. Current medication. Serious illnesses or operations. General medical or surgical investigations.

7. Family history. Parents' state of health. If they are not alive, the age at and cause of death. Note if the patient is adopted, fostered or in care.

8. Specific points related to each method. Specific questions need to be asked to assess the patient's suitability for certain methods, and are dealt with in the appropriate chapters.

Examination

The nature and extent of examinations carried out will be determined by the method requested, by the presence of symptoms, by previous examination findings recorded in the notes, and by clinic or practice screening policies.

Basic examinations include:

1. Observation on general health.

2. Recording weight and blood pressure.

3. Examination of the abdomen, breasts and pelvis including cervical smear when appropriate.

4. Additional examinations which may be indicated for certain methods.

Provision of information to the patient

Time spent on providing relevant and adequate information will pay dividends. It is important to:

1. Use language she can understand. The words 'ovulation, cycle, erosion, smear' are but a few of the terms we use which are unfamiliar to many patients. Where necessary simple alternative words should be used and explained.

2. Concentrate on essentials and explain them clearly.

3. Where a range of options is being considered, show her each contraceptive and how it works. Diagrams and plastic models may be used but, while helpful to some, they can be totally mystifying to others.

    a. IUD. Give her the opportunity to handle it and show her how it is inserted. Then hold the device between the first finger and the thumb so that it lies within the space between them which can be likened to the uterine cavity. Cup the other hand around the threads to represent the vagina with the threads lying in the upper part.

    b. Cap. The insertion of the cap and its position over the cervix can also be shown using the hands.

    c. Sheaths. Do not assume that patients are familiar with their use. Explain clearly how to use them.

    d. Spermicides. Show them and explain their use.

    e. Natural family planning. Have a fertility thermometer, temperature and Billings charts on show. If she chooses this method it may be advisable to refer her to an expert teacher.

    f. Sterilisation. Provide diagrams and leaflets to reinforce the points made in discussion because many people have only the vaguest idea of what sterilisation entails.

    g. Oral contraception. The simplicity of pill-taking regimens can best be illustrated by using a packet of pills.

4. When she has finally chosen the method she wants to use, it will

have to be thoroughly explained to her and she should be told what she will experience and what differences in herself she should expect.

Continuing support and follow-up

At each return visit it is important to establish that the contraceptive is being used correctly, that no contraindication to its use has developed and that the patient is still happy with her choice. She should be given an opportunity to discuss any problems or ask questions. Symptoms which seem unimportant to the doctor may be troublesome and significant to the patient. Reassurance should be given only if it is fully justified and accompanied by a comprehensible explanation. It may be necessary to arrange for treatment, early review or a change of method. Before leaving, the patient should always know when to return.

Minor side effects

Much can be done to keep these at a minimum by proper history taking, examination, follow-up and judicious choice of method. The patient's fears may be concerned with adverse effects not harmful to health in the long term but which may be interfering seriously with the quality of the individual's life. The object of every family planning consultation should be to ensure that the patient is using the right method for her and that its use is as trouble-free and risk-free as possible.

Checklist

The main criteria which must be fulfilled for all patients can be checked by considering the following questions.

1. How important is it that the woman should not become pregnant?
2. Does the method suit the couple's requirements?
3. Are there any contraindications to its use?
4. Does the patient know enough about it?
5. Is the patient likely to use it correctly?
6. Will it be convenient to use?
7. Is the method and its follow-up routine going to be easily available?
8. Is this the method the individual or couple really want to use and does the final choice of contraceptive represent a reasonable balance of safety and effectiveness?

GROUPS WITH SPECIAL NEEDS

Young people

Age in years is less significant than maturity, attitude, sexual experience and intention and the stability of the individual's background. In assessing contraceptive needs and planning the future care of a young girl, it is important to have answers to the following questions:

1. Has she had intercourse and, if so, when did she start?

2. If not, does she understand the significance of what she now wishes and intends to do?

3. Has she or her boyfriend ever used contraception?

4. Do her parents know she is seeking contraception and, if not, does she plan to tell them?

5. Is she willing to have letters sent to her home address and, if not, will she provide a contact address? If she is unwilling to be contacted make sure that this is clearly recorded in her notes and on laboratory forms. It is often safer to omit the address altogether from the latter. She will then need to contact the clinic herself for the results of any investigations.

6. Does she agree that her family doctor may be informed about her choice of contraception? She should understand that such information may be very important to him in her future medical care.

A memorandum of guidance on family planning services for young people has been issued by the DHSS (February 1981). It expresses a hope that doctors and other professional workers will always seek to persuade the child under the age of consent to involve a parent or guardian if contraceptive advice is sought, but acknowledges the need for confidentiality. It concludes that a decision whether or not to prescribe contraception must be for the clinical judgement of the doctor (Ch. 14). For some it is unrealistic to urge discussion with parents or guardian and irresponsible to withhold contraception.

Some young people are grateful for a discussion which leaves them with the assurance not to embark on intercourse just yet. These girls should always be given a further appointment for they often require continuing support.

For those who require contraception:

1. A low-dose oestrogen/progestogen pill is probably best for a girl whose periods are well established, provided she can be relied upon to take it regularly.

2. Progestogen-only pills are less acceptable to this age group since

they often produce cycle irregularity and are not as effective as combined preparations.

3. A diaphragm may be suitable for some girls but at this age they are often too poorly motivated to use it reliably.

4. IUDs are best avoided except in special circumstances (Ch. 7).

5. Condoms are very good contraceptives if the partner can be persuaded to use them properly and consistently. Some girls prefer the extra assurance provided by using pessaries as well.

6. The vaginal contraceptive sponge may appeal to this age group since it will be available across the counter, but its efficacy, shown in clinical trials to be in the range of that of other vaginal contraceptives is yet to be proven, when in general use.

Older women

The thought of an unplanned pregnancy usually holds particular fears for women over the age of 40. Some women of this age have never sought contraceptive advice from a doctor and only a crisis situation such as a delayed period brings them to seek help.

1. Sterilisation of the woman or her partner is commonly chosen, though patients should always be informed of the other options available to them.

2. The low-dose combined or triphasic pill may still be prescribed for healthy nonsmokers under 45. The progestogen-only pill is particularly useful for the older woman who smokes.

3. An IUD may be inserted.

4. Those who have used the cap or sheath for years may be happy to continue to do so but these methods do not usually appeal to couples who have never used them before.

Contraception should be continued for 1 year after the last menstrual period if the woman is over 50 and for 2 years if she is under 50.

Mentally retarded patients

Some mentally retarded patients have close personal and sexual relationships of a stable character. Others may be vulnerable to exploitation or to casual sexual contact. Patients can sometimes be misleading about the risk of pregnancy which they run. Partners, parents, a nurse or social worker may be helpful in assessing how great the risk of pregnancy really is and their support may be needed for the reliable use of contraception.

Subnormal patients are often fearful of examinations and special care should be taken not to frighten or hurt them. If they are capable of understanding, they should be told in simple terms that the sexual feelings and emotions which they may experience are perfectly normal. An explanation of how conception may occur and why contraception is advised should be attempted.

1. If it is taken or given correctly, the combined pill will provide the most effective contraception. An everyday (ED) preparation may be best.

2. If insertion is tolerated, then the IUD is ideal for those whose pill taking might be unreliable.

3. The progestogen-only pill may be useful with its everyday routine.

4. Injectables are indicated in some cases.

5. Sterilisation may be the answer but is fraught with legal problems (Ch. 14).

Physically handicapped patients

The method prescribed must be one which the patient is physically capable of using. Ideally it should not create extra difficulties or risks. If menstruation is already difficult to deal with, the IUD may be ill-advised and the oral contraceptive pill with its reduced blood loss may be better. However, the increased risk of thrombosis in an immobile woman may make the pill unacceptable.

The patient's partner can be taught to insert a cap for her if necessary.

The Association to Aid the Sexual and Personal Relationships of the Disabled (SPOD) provides an advisory and counselling service for disabled people with sexual difficulties and is a valuable source of information for those who work with them (p. 324).

Twelve leaflets covering all aspects of contraception are now available in braille, free from Family Planning Information Services. They include information on postpartum and postcoital contraception and a leaflet for teenagers.

After delivery

It is impossible to predict when ovulation will return after delivery but it has not been recorded earlier than the 33rd day. Theoretically, therefore, there is no need to use contraception if intercourse takes place during the first month postpartum but, to avoid confusion and possible default, it is wise to advise disadvantaged or ill-motivated patients to do so when intercourse is resumed. Condoms are generally acceptable until other methods can be started and relied upon.

1. The pill may be started without waiting for a menstrual period. If the patient is breast feeding a progestogen-only pill is indicated (pp. 95, 101).

2. The IUD may be inserted within 72 hours of delivery but the expulsion rate is then high. Insertion is often delayed until the first postnatal visit (p. 140).

3. Cap fitting should be delayed until the first postnatal visit.

4. It is recommended that the use of injectables be delayed until six weeks postpartum (p. 118).

5. Sterilisation (pp. 214, 217).

After termination of pregnancy

Before termination of pregnancy is carried out future contraception should be discussed, the method decided upon and arrangements made for follow up (p. 238). If sterilisation is requested and considered appropriate this may be carried out at the time of operation or arranged for a later date (p. 217).

1. The pill may be started immediately, even before the patient leaves the hospital.

2. The IUD may be inserted at the time of a first trimester termination or at the subsequent follow up visit.

3. Cap fitting should be delayed until the first follow up visit.

4. An injectable may be given on the day of termination of pregnancy. Side effects are the same as for interval injectable use.

Patients of a different culture

Cultural backgrounds may influence attitudes towards fertility and the acceptability of certain methods of contraception. It is impossible to generalise since national attitudes tend to merge in mixed communities. It is important to realise that some methods may be unacceptable because of religious or cultural beliefs or taboos and the patient should always be given an opportunity to say so, e.g. some West Indian women like the IUD because of a strong cultural belief that 'a good bleed gets the badness out of the body'. On the other hand, for Moslem women who are forbidden to handle food while 'bleeding', the coil may be severely limiting and the restriction imposed quite unacceptable, if the woman also has menorrhagia or intermenstrual spotting.

Poorly motivated patients

There are some patients who might clearly benefit from good family planning advice but who will not attend clinics or surgeries for this purpose. They include the mothers of large families, of children in care or with very little care, those with haphazard personal and social relationships, in trouble with the law, and with substandard living conditions. These women may be brought to the clinic or the surgery by the health visitor or social worker. They may require to be visited at home by a member of the primary care team or the domiciliary family planning team and diligent follow-up is essential. Methods such as the IUD or the injectable which do not require the patient's co-operation are often best. For those whose family is complete, sterilisation may be indicated but this should only be carried out after careful counselling.

FACTORS INFLUENCING THE DOCTOR AND NURSE

Both doctor and nurse can have considerable influence on a patient's choice of method. Personal views and beliefs, as opposed to knowledge and expertise, should not be allowed to colour the advice they give. If there is a moral conflict then only exceptionally should the adviser's view prevail over that of the patient. When strongly held ethical beliefs do not allow unbiased consideration of a patient's request she should be referred elsewhere.

When helping a patient to choose a contraceptive, convenience, safety, effectiveness and acceptability must all be taken into account. Any objection by the patient to a method, however irrational it may seem, will lessen the likelihood of that method being used to good effect. It may be better to advise a theoretically less effective method which, with good motivation, could have a higher degree of success over a long period of time.

Both doctors and nurses are influenced by what they know of the various contraceptive techniques — of their effectiveness and of the mortality and morbidity associated with their use.

Effectiveness

An indication of the degree of effectiveness which may be achieved by well motivated couples is shown in Table 2.1. The data come from the Oxford Family Planning Association (Oxford/FPA) Contraceptive Study and refer to 17 000 married women who were aged 25–30 at recruitment to the study and who have been observed for 9 years or

Table 2.1 Failure rates for different methods of contraception. (Oxford/FPA Contraceptive Study: married women aged 25–30) Vessey et al 1982)

Method	Failure rate per 100 women years
Combined pills	
50 μg oestrogen	0.16
<50 μg oestrogen	0.27
Progestogen-only pill	1.2
IUD — average for all types	1.5
Diaphragm	1.9
Condom	3.6
Chemicals alone	11.9
Rhythm	15.5
Coitus interruptus	6.7
Sterilisation	
male	0.02
female	0.13

more. About 40% of these women were from social class I or II and were thought to have more positive attitudes to health than average. The effectiveness of all methods, especially barrier methods and IUDs was found to be highest in the upper age groups and in those who had used them longest. For those less dedicated in their use of contraception the risk of failure will be greater than these figures imply. It is important to assess the patient's individual potential for success or failure when using various methods rather than to assume that the success rates achieved in this study can be applied to all. The enthusiasm and teaching ability of the doctor and nurse may enhance the success of patient-dependent methods in many cases.

Mortality and morbidity

It is not possible to enjoy the effectiveness and benefits of modern contraception without some disadvantages. Many patients will not use the pill, the IUD, or accept sterilisation because they are afraid of the risks. These risks have to be put in perspective and must be weighed against the risk of pregnancy when no contraceptive or a less effective one is used. It should be borne in mind that without contraception 80% of sexually active women will conceive within one year.

It is therefore not a question of choosing between a dangerous option and a completely safe one. It is more a matter of weighing up the advantages and risks of one contraceptive against another and the disadvantages and risks that might be associated with pregnancy in each individual case. In this the doctor and nurse have a very important role to play.

Mortality

Fortunately death directly attributed to a contraceptive is extremely rare. Maternal mortality is approximately 13 per 100 000 pregnancies. Both the Oxford/FPA Study and that of the Royal College of General Practitioners have shown that it is almost always safer to use any method of contraception than to have a baby. Tietze & Lewit (1979) in a computer-based study of life risks associated with reversible contraceptive methods in America, reached the same conclusion (Table 2.2), the exception being women in their late 30s and 40s who took the pill and who smoked. At all ages the lowest mortality rate was achieved by the use of a barrier method with the back-up of early abortion in case of failure.

Table 2.2 Birth-related, method-related and total deaths per 100 000 women per year by age (Tietze & Lewit 1979)

Regimen	Age					
	15–19	20–24	25–29	30–34	35–39	40–44
No contraception birth-related only	5.3	5.8	7.2	12.7	20.8	21.6
Abortion method-related only	1.0	1.9	2.4	2.3	2.9	1.7
Pill-non-smokers birth-related (0.1–0.6)* + method-related	0.7	1.3	1.8	3.4	9.7	18.1
Pill — smokers birth-related (0.1–0.6)* + method-related	2.2	4.4	6.3	12.2	31.9	61.3
IUD birth-related (0.1–0.6)* + method-related	0.9	1.0	1.2	1.4	2.0	1.8
Barrier birth-related only*	1.1	1.5	1.9	3.3	5.0	4.0
Barrier + abortion method-related	0.1	0.3	0.4	0.4	0.4	0.2

*Deaths occurring as a result of failure of the method

Morbidity

Complications and morbidity associated with different methods of contraception are discussed in detail in the appropriate chapters. Non-obstetrical morbidity associated with different reversible methods of contraception is summarised in Table 2.3.

Table 2.3 Non-obstetrical morbidity associated with different reversible methods of contraception. Rates of hospital admission per 100 000 women of childbearing age per annum (Vessey 1982)

Condition	Pill	Diaphragm	IUD
Menorrhagia	260	300	440
Anaemia	40	60	90
Benign lesions of breast	120	280	
Functional ovarian cysts	85	120	
Venous thromboembolism	90	20	
Stroke	45	10	
Acute myocardial infarction			
(non-fatal) ages 30–39	6	2	
40–44	55	10	
Cholelithiasis	180	120	
Hepatocellular adenoma	3	0	
Cervical erosion	540	250	
Pelvic inflammatory disease	55		200
Uterine perforation	—		50

REFERENCES

DHSS February 1981 Health services management. Family planning services for young people. DHSS: HN (81) 5 LASSL (81) 2

Miller D L, Farmer R D T 1982 (eds) Epidemiology of diseases. Blackwell, Oxford

Royal College of General Practitioners 1977 Mortality among oral contraceptive users. Lancet ii: 727–731

Tietze C, Lewit S 1979 Life risks associated with reversible methods of fertility regulation. International Journal of Gynaecology and Obstetrics 1978–79; 16.6: 456–459

Vessey M, Lawless M, Yeates D 1982 Efficacy of different contraceptive methods. Lancet i: 841–2

Hormonal contraception: development and mode of action

Hormonal contraceptive preparations
  Combined oestrogen/progestogen pills
    Monophasic (fixed dose) pills
    Biphasic pills
    Triphasic pills
  Progestogen-only pills
  Injectables

Mode of action of oestrogen
  Hypothalamus and anterior pituitary gland
  Ovary
  Endometrium

Mode of action of progesterone
Mode of action of progestogens
  Hypothalamus and pituitary gland
  Ovary

Endometrium
Cervical mucus
Fallopian tube

Mode of action of hormonal contraceptives
  Combined oestrogen/progestogen pills
  Progestogen-only pills
  Injectables

Dose and potency of steroid hormones
  Oestrogens
  Progestogens
    Potency
  Progestogen or oestrogen dominance of
    combined pills

In 1921 Haberlandt was the first scientist to indicate that extracts from the ovaries of pregnant animals might be used as oral contraceptives. In 1937 Kurzrok noted that ovulation was inhibited during treatment for dysmenorrhoea with ovarian oestrone, and suggested that this hormone might be of value in fertility control. It was only in the 1950s, when potent, orally active progestogens (first norethynodrel and then norethisterone) became available, that an oral contraceptive pill became possible. The research chemists chiefly responsible were Russell Marker, who first produced progesterone from diosgenin extracted from the Dioscorea plant, George Rosenkranz, Carl Djerassi and Frank Colton (Djerassi 1979).

The first pills were thought to contain only progestogen and gave good cycle control. When purified preparations were tried cycle control deteriorated. The impurity had been mestranol, and when this oestrogen was restored to the tablets the combined pill Enovid (norethynodrel plus mestranol) was created. Successful trials with this preparation started in Puerto Rico in 1956.

The pill was approved for use in America in 1959 and in Britain two years later. For years its use appeared to be associated with only minor side-effects which were acceptable to most women in return for its high effectiveness. Cardiovascular problems, related to the dose of oestrogen, came to light in 1969. Later progestogen was also implicated. Over the years the composition of the pill has changed markedly — the total dose of steroid has been reduced; ethinyl oestradiol has largely replaced mestranol; some progestogens have been abandoned and new ones introduced.

Further attempts to reduce the total steroid content of the pill have led to the introduction of phased preparations (Ch. 4).

At least 150 million women throughout the world have used the pill since it first became available. At present about 65 million women rely on it and in the United Kingdom it is the most popular method of fertility control, being used by almost 3 million women.

In an effort to avoid the metabolic side effects of oestrogen, progestogen-only pills were introduced in the 1960s using derivatives of 19-nortestosterone and of 17-acetoxyprogesterone. The latter were soon abandoned.

Progestogens by intramuscular injection have been used for the past 20 years (Ch. 6).

Progesterone-containing IUDs, e.g. the Progestasert, enjoyed a brief vogue. Although they fell into disrepute, they had great potential and seem likely to reappear for routine clinic use.

Research is currently taking place into delivery of steroid hormones by vaginal rings, silastic implants and biodegradable microspheres.

HORMONAL CONTRACEPTIVE PREPARATIONS

Combined oestrogen/progestogen pills

Monophasic (fixed dose) pills

Twenty-nine combined oral contraceptive preparations are listed in Tables 3.1 and 3.2. Each contains 1 of 6 progestogens combined with 1 of 2 oestrogens. Each packet contains 21 active pills except Minilyn, in which there are 22. Each course of pills is followed by either 7 pill-free days (6 Minilyn-free days) or by 7 placebo tablets which permit exogenous steroid levels to decline and withdrawal bleeding to occur. This tablet-free interval mimics the physiological decline in steroid hormones that occurs in the normal menstrual cycle.

Preparations containing 35 or 30 micrograms of ethinyloestradiol are now the most popular varieties. One pill, Loestrin 20, contains only 20 micrograms of oestrogen. This low dose creates problems with cycle

Table 3.1 Combined pills containing 50 micrograms of oestrogen and available in Britain

Name of pill	Dose of progestogen in milligrams	Remarks
Norethisterone group Ortho-Novin 1/50/Norinyl-1	1.0	1. In both, oestrogen is mestranol.*
Norethisterone acetate group Anovlar 21 Gynovlar 21 Norlestrin	4.0 3.0 2.5	
Orlest 21/Minovlar Minovlar ED	1.0	Minovlar ED includes 7 placebo tablets
Ethynodiol diacetate group Ovulen 50	1.0	
Lynestrenol group Minilyn	2.5	This is the only 22 tablet regimen. It is followed by 6 tablet-free days
Levonorgestrel group Ovran/Eugynon 50	0.25	Eugynon 50 also contains 0.25 mg of inactive isomer of progestogen

*The oestrogen in the other pills is ethinyloestradiol.

Table 3.2 Low-dose combined pills containing less than 50 micrograms of oestrogen and available in Britain

Name of pill	Dose of oestrogen (ethinyloestradiol) in micrograms	Dose of progestogen in milligrams	Remarks
Norethisterone group			
Norimin/Neocon 1/35	35	1.0	
Binovum	35 35	0.5 for 7 days 1.0 for 14 days	Biphasic preparation
Trinovum	35 35 35	0.5 for 7 days 0.75 for 7 days 1.0 for 7 days	Triphasic preparation
Brevinor/Ovysmen	35	0.5	
Norethisterone acetate group			
Loestrin 20	20	1.0	
Loestrin 30	30	1.5	
Ethynodiol diacetate group			
Conova 30	30	2.0	
Levonorgestrel group			
Ovran 30/Eugynon 30	30	0.25	
Ovranette/Microgynon 30	30	0.15	
Trinordiol/Logynon	30 40 30	0.05 for 6 days 0.075 for 5 days 0.125 for 10 days	Triphasic preparation
Logynon ED			Logynon ED has 7 placebo tablets
Desogestrel group			
Marvelon	30	0.15	

control and has been reported as being less effective in preventing pregnancy.

Biphasic pills

There is only one biphasic preparation (Binovum) in current use (Table 3.2). The rationale for the biphasic formulation is basically similar to that of the triphasic regimen (see below). It differs significantly from previous 'biphasic' preparations known as 'sequentials' in that unopposed oestrogen is never administered during the cycle.

Each course of 21 pills is followed by 7 pill-free days.

Triphasic pills

There are two formulations but three brand names are used (Table 3.2). The object of this regimen is to mimic the well-known and characteristic fluctuations in oestrogen and progesterone which occur during the menstrual cycle. In practice, a histologically more normal-looking endometrium develops, giving adequate cycle control, despite using relatively low total quantities of steroids. The latter should reduce long-term risks, though whether the phased administration is of itself beneficial remains to be proved.

Each course of pills is followed by 7 pill-free days.

Progestogen-only pills

These pills, sometimes called 'minipills', are oestrogen-free. A small dose of progestogen is taken daily on a continuous basis. The six preparations currently available are listed in Table 3.3.

Table 3.3 Progestogen-only pills

Name of pill	Progestogen	Dose in micrograms
Femulen	Ethynodiol diacetate	500
Noriday	Norethisterone	350
Micronor	Norethisterone	350
Neogest	Levonorgestrel	37.5*
Microval	Levonorgestrel	30
Norgeston	Levonorgestrel	30

*Plus 37.5 μg of inactive isomer

Injectables

Depot injections of progestogen are given intramuscularly at regular intervals. Depot medroxyprogesterone acetate (DMPA or Depo-Provera) is usually injected intramuscularly in a dose of 150 mg every 3 months. The dose of norethisterone oenanthate (NET OEN, or Noristerat) is 200 mg every 8 weeks (Ch. 6).

MODE OF ACTION OF OESTROGEN

Hypothalamus and anterior pituitary gland

Ethinyloestradiol and mestranol are potent inhibitors of ovulation. They act by reducing the levels and rate of pulsatile release of gonadotrophin-releasing hormone secreted from the hypothalamus to the pituitary gland via its portal blood system. The early cycle increase in follicle stimulating hormone (FSH) does not occur and, in turn, there is no rise in plasma oestradiol from the ripening follicle of either ovary to stimulate a midcycle surge of luteinising hormone (LH). Because of the absence of a developed follicle and LH surge, ovulation does not occur.

Ovary

The reduction of FSH levels leads to reduced ovarian activity and failure of follicular maturation. The production of endogenous ovarian steroids is therefore diminished, though there may be a rise in some women during the tablet-free interval of each pill cycle.

Endometrium

The administration of oestrogen produces proliferative endometrium. This builds up from the basal cells remaining after menstruation both by an increase in the volume of the endometrial glands and stroma, and by growth of blood vessels, mainly the branches of the spiral arterioles.

MODE OF ACTION OF PROGESTERONE

Progesterone is a naturally occurring steroid hormone produced by the corpus luteum of the ovary, by the adrenal cortex and by the placenta during pregnancy. Its two main functions relevant to this discussion are:

1. The production of the secretory endometrium which develops after ovulation.
2. The production of a decidual reaction during pregnancy.

Progesterone itself is not used therapeutically in contraceptive preparations because it does not produce sustained high blood levels after oral administration. However, some of its actions have a contraceptive function and these have been imitated to some extent by the synthetic progestogens which form one component of the combined pill.

MODE OF ACTION OF PROGESTOGENS

Synthetic progestogens do not have all the effects of endogenous progesterone. They act on:

Hypothalamus and pituitary gland

Progestogens interfere with the pulsatile release of gonadotrophin-releasing hormone and also influence pituitary production of LH. Suppression of ovulation may occur but with small doses is not a constant effect.

Large doses of progestogens, given by intramuscular injection, lead to suppression of the pituitary gonadotrophin production and therefore to suppression of ovulation.

Ovary

Small doses of 19-nortestosterone derivatives impair or suppress luteal function. This does not occur with the same dose of medroxy-progesterone derivatives.

Endometrium

The effect of synthetic progestogens on the endometrium varies with the type of progestogen and the duration of its administration. The 19-nortestosterone derivatives used in oral contraceptive pills alter the development of the endometrial stroma and glands and cause inadequate secretory activity. The stroma becomes excessively vascular and oedematous and has the cellular appearance of decidua, contrasting with the glands which are atrophic with an absence of secretion. It is believed that this unprepared endometrium resists implantation and fails to maintain the blastocyst.

Cervical mucus

Progestogens make cervical mucus more tenacious and more resistant to sperm penetration. They interfere with the typical oestrogen-induced characteristics of cervical mucus, which normally occur at midcycle and which are essential for the activity of spermatozoa.

Fallopian tube

Synthetic gestogens affect the tubal epithelium and so interfere with the transport of the ovum and sperm within the tube.

MODE OF ACTION OF HORMONAL CONTRACEPTIVES

After considering the actions of both oestrogens and progestogens the mode of action of hormonal contraceptives can be summarised as follows (Table 3.4):

Combined oestrogen/progestogen pills

1. Inhibition of ovulation. There are two mechanisms: a. absence of follicular ripening, b. absence of LH surge.
2. Alteration of the endometrium: the usual development does not occur. An atrophic endometrium is produced with microtubular glands and a fibroblastic condensation of the stroma. This reduces the risk of implantation, should breakthrough ovulation occur. These changes are less marked histologically with phased pills.
3. Alteration of cervical mucus: this renders it hostile to sperm penetration.
4. Possible direct effect on the ovary and on the fallopian tubes: these are of doubtful importance.

Progestogen-only pills

1. Changes in the endometrium making it unreceptive to the implantation of the fertilised ovum.
2. Alteration in the cervical mucus rendering it hostile to sperm penetration.
3. Decreased tubal function.
4. Ovulation suppression in some women in some cycles.
The first two are believed to be the main contraceptive effects.

Table 3.4 Mechanism of action of hormonal contraception (the more +s means the greater the effect)

	Combined pills	Progestogen-only pills	Injectable progestogens
1. Reduced FSH, therefore follicles prevented from ripening and ovum from maturing	++++	+	+++
LH surge stopped so no ovulation	++++	++	+++
2. Endometrium less suitable for implantation	+++	++(+)	+++
3. Cervical mucus changed into a barrier to sperm	+++	+++	+++
4. Impairment of tubal transport	+	+	+
Expected pregnancy rate per 100 women using the method for one year (compare use of NO METHOD = 80–90)	0.1–1.0	0.5–4.0	0.1–1.0

Injectables

1. Inhibition of ovulation

   a. DMPA: peak serum levels, which occur within 24 hours and are followed by an effective plateau for 2 or 3 months, inhibit pituitary secretion of LH and FSH, prevent follicular growth and abolish the cyclical surge of gonadotrophins.

   b. NET OEN: high serum levels normally inhibit ovulation for 60 days but 'breakthrough' ovulation may then occur.

2. Suppression of endometrium

   a. DMPA: binding to progesterone receptors in an oestrogen primed endometrium inhibits proliferation. The endometrium may become 'atrophic' but usually recovers about 90 days after 150 mg dose.

   b. NET OEN: variable endometrial effects, not so marked as with DMPA.

3. Hostile changes in cervical mucus. Both preparations render the cervical mucus hostile to sperm penetration.

4. Alteration in tubal function. Both preparations have this contraceptive effect.

DOSE AND POTENCY OF STEROID HORMONES

Oestrogens

Mestranol is the 3-methyl-ether derivative of ethinyloestradiol, and the latter is its biologically active metabolite in the body. About 80% conversion occurs, so mestranol is less potent and is now used in only two medium-dose formulations in the UK.

Both oestrogens affect coagulation factors in such a way as to promote intravascular coagulation, though there is evidence of compensatory increased fibrinolytic activity in most pill-users. These changes are less marked the lower the dose given. As the epidemiology of venous thromboembolism shows the same correlation with dosage, oral contraceptive pills in Britain do not contain more than 50 μg of oestrogen.

Both oestrogens also tend to raise plasma lipid levels, including high density lipoprotein-cholesterol (HDL-C) — in this respect having an effect opposite in direction to that observed with increasing potency of administered progestogens (p. 51). The resultant effect on HDL-C of the two hormones in combined pills can be used as a measure of relative oestrogen dominance, but whether *raised* levels are truly beneficial if achieved in this way is unproven.

Progestogens

Those in current use are norethisterone and its acetate, lynestrenol, ethynodiol diacetate, norgestrel chiefly as levonorgestrel, and desogestrel. All are derivatives of 19-nortestosterone and may be divided into two groups.

Group 1 (norethisterone, norethisterone acetate, lynestrenol, ethynodiol diacetate). Since norethisterone is the chief metabolite of the other three, all four can be considered as having approximately the same potency, weight for weight.

Group 2 dl-norgestrel is a racemic mixture of a d-isomer and l-isomer. Only the former is biologically active and since it causes a solution of the substance to deviate light in the opposite direction (i.e. to the left) it is known as levonorgestrel.

Desogestrel is a prodrug requiring conversion to its active metabolite (the 11 methylene derivative of levonorgestrel) before acquiring biological activity. Its potency is very similar to that of levonorgestrel but it is less androgenic.

Both levonorgestrel and desogestrel are of equivalent strong potency.

Potency

Potency describes the comparative amounts of two or more drugs required to produce the same response and it is important to understand what the term means when applied to the action of oral contraceptive pills. Steroids act on many tissues in the body — the brain, pituitary, breast, cervix, endometrium etc. The effect they produce depends on many factors such as whether the tissues have been oestrogen primed, the mode of administration, whether the drug is given continuously or intermittently, etc. All these factors must be taken into account when trying to assess potency and various tests have been devised to compare the potency of different progestogens. Those in animal studies cannot be extrapolated to man. In humans the most reliable tests are those relating to the endometrium.

1. The delay of menses test (Swyer 1982).
2. The subnuclear vacuolation test (Dickey & Stone 1976).

From these two tests it has been calculated that levonorgestrel is approximately five times as potent as norethisterone, norethisterone acetate, ethynodiol diacetate and lynestrenol. At the moment that is the best estimate that one can offer.

It is important to remember that this potency has been calculated on the progestogen effect on the endometrium. It does not necessarily mean that levonorgestrel is five times as potent as norethisterone in its effect on other tissues in the body. For example, there is at present no test by which the effect of different progestogens on the breast tissue can be measured and compared.

Observations on the effect of a 1 mg dose of each progestogen (combined with a constant dose of oestrogen) on plasma HDL-C levels give support for the view that progestogens in group 2 are much more potent than those in group 1 (Briggs 1979).

Progestogen or oestrogen dominance of combined pills

In assessing the relative progestogen or oestrogen dominance of combined pills, we must take into account the dose of both oestrogen and progestogen, and the potency of the progestogen. By doing this it is possible to draw up a rough classification of the combined pills available in the UK along a scale of relative progestogen/oestrogen dominance (Tables 3.5 and 3.6). However, it must be stressed again that there is no one exact figure which can be set against a progestogen quantifying its potency in the human. Thus data for making comparisons between the two major groups of pills are limited and not fully constant.

Marvelon produces a greater elevation of sex hormone binding globulin than comparable levonorgestrel products and slightly raises levels of HDL-C. It is certainly more oestrogen dominant clinically than Ovranette and Microgynon (e.g. in its effect on acne), but whether other effects, especially that on HDL-C levels, will prove to be beneficial in the long run from the point of view of cardiovascular disease remains to be proved.

Table 3.5 50 microgram oestrogen pills listed approximately in order of decreasing relative progestogen dominance and of increasing oestrogen dominance

Norethisterone group	Levonorgestrel group
Anovlar 21	
———	
Gynovlar 21	
———	
Norlestrin	
Minilyn	
——— ◄———————————	Ovran/Eugynon 50
Norinyl-1/Ortho-Novin 1/50	
Ovulen 50	
Minovlar/Orlest	

Table 3.6 Low dose oestrogen pills listed approximately in order of decreasing relative progestogen dominance and of increasing oestrogen dominance

Norethisterone group	Levonorgestrel group
Conova 30	Eugynon 30/Ovran 30
— —	
Loestrin 20*	
———	
Loestrin 30	
———	
Norimin/Neocon 1/35	
———	
Binovum	
◄———————————	Microgynon 30/Ovranette
Trinovum	Marvelon
———	Logynon/Trinordiol
Brevinor Ovysmen	———

*Assigned to this point because it is so oestrogen deficient

(See Ch. 4 for further discussions of the implications of Tables 3.5 and 3.6 for pill prescribing.)

REFERENCES

Briggs M H 1979 Recent biological studies in relation to low-dose hormonal contraceptives. British Journal of Family Planning 5: 25–28
Dickey R P, Stone S C 1976 Progestational potency of oral contraceptives. Obstetrics & Gynaecology 47: 106–12
Djerassi C 1979 The chemical history of the pill. In: The politics of contraception. W W Norton & Co. New York, p. 227–55
Kurzrok R 1937 Prospects for hormonal sterilization. Journal of Contraception 2: 27–29
Swyer G I M 1982 Potency of progestogens in oral contraceptives — further delay of menses data. Contraception 26: 23–27

Combined oral contraceptive pills

Although this method is easy, can provide maximum protection from pregnancy and has many beneficial side effects, it is neither suitable for nor acceptable to all couples.

Anxiety about adverse effects and possible long-term consequences makes it necessary for all those who prescribe the pill to:

1. Form their own opinions based on scrutiny of published work.
2. Keep regularly updated.
3. Prescribe carefully after discussing anxieties, risks and benefits.
4. Reassure patients when appropriate, but leave final decisions to them.
5. Supervise follow-up conscientiously.

Two points have to be considered whenever the pill is to be prescribed: the particular pill and the individual woman.

TYPES OF PILLS

All combined oral contraceptive pills (OCs; pills) available in the UK are listed in Chapter 3 (pp. 42, 43), and classified into four groups for easy reference in Table 4.1.

MODE OF ACTION

Combined OCs have three main sites of action, which may be summarised as follows:
1. Inhibition of ovulation (two mechanisms).
2. Suppression of the endometrium.
3. Alteration in the composition of cervical mucus.
For details see Chapter 3, p. 47.

EFFECTIVENESS

Provided the pill is taken correctly and consistently (pp. 77, 78, 80), is absorbed normally and its metabolism is not affected by interaction (p. 81) with other medication, its reliability is nearly 100%. In practice the failure rate is 0.1 to 1 per 100 woman years, depending on the population studied.

It is probable that many more errors in tablet-taking occur than are reported. Detailed investigation nearly always reveals the reason for any unexpected pregnancy.

Careful teaching of the woman and sometimes her partner is essential. Written instructions help, but not all women bother to read them and a few are unable to do so.

Table 4.1 Combined oestrogen/progestogen pills

	Preparation	Progestogen	Dose	Oestrogen	Dose
			mg		µg
Pills containing 50 µg oestrogen	Norinyl – 1	Norethisterone	1.0	Mestranol	50
	Ortho-Novin 1/50	Norethisterone	1.0	Mestranol	50
	Anovlar 21	Norethisterone acetate	4.0	Ethinyloestradiol	50
	Gynovlar 21	Norethisterone acetate	3.0	Ethinyloestradiol	50
	Norlestrin	Norethisterone acetate	2.5	Ethinyloestradiol	50
	Minovlar	Norethisterone acetate	1.0	Ethinyloestradiol	50
	Minovlar ED	Norethisterone acetate	1.0	Ethinyloestradiol	50
	Orlest 21	Norethisterone acetate	1.0	Ethinyloestradiol	50
	Minilyn	Lynestrenol	2.5	Ethinyloestradiol	50
	Ovulen 50	Ethynodiol diacetate	1.0	Ethinyloestradiol	50
	Eugynon 50	Levonorgestrel	0.25*	Ethinyloestradiol	50
	Ovran	Levonorgestrel	0.25	Ethinyloestradiol	50
Pills containing < 50 µg oestrogen	Brevinor	Norethisterone	0.5	Ethinyloestradiol	35
	Oysmen	Norethisterone	0.5	Ethinyloestradiol	35
	Norimin	Norethisterone	1.0	Ethinyloestradiol	35
	Neocon 1/35	Norethisterone	1.0	Ethinyloestradiol	35
	Conova 30	Ethynodiol diacetate	2.0	Ethinyloestradiol	30
	Eugynon 30	Levonorgestrel	0.25	Ethinyloestradiol	30
	Microgynon 30	Levonorgestrel	0.15	Ethinyloestradiol	30
	Ovran 30	Levonorgestrel	0.25	Ethinyloestradiol	30
	Ovranette	Levonorgestrel	0.15	Ethinyloestradiol	30
	Marvelon	Desogestrel	0.15	Ethinyloestradiol	30
	Loestrin 30	Norethisterone acetate	1.5	Ethinyloestradiol	30
	Loestrin 20	Norethisterone acetate	1.0	Ethinyloestradiol	20

*Contains 250 µg of inactive isomer.

Table 4.1 (Cont.)

Preparation	Progestogen	Dose mg	Oestrogen	Dose µg
Biphasic pills				
Binovum	a) 7 pills contain norethisterone	0.5	Ethinyloestradiol	35
	b) 14 pills contain norethisterone	1.0	Ethinyloestradiol	35
Triphasic pills				
Trinordiol / Logynon	a) First 6 pills contain levonorgestrel	0.05	Ethinyloestradiol	30
	b) Next 5 pills contain levonorgestrel	0.075	Ethinyloestradiol	40
	c) Remaining 10 contain levonorgestrel	0.125	Ethinyloestradiol	30
Logynon ED	As above but preceded by 7 placebo tablets each containing inert lactose			
Trinovum	a) 7 pills contain norethisterone	0.5	Ethinyloestradiol	35
	b) 7 pills contain norethisterone	0.75	Ethinyloestradiol	35
	c) 7 pills contain norethisterone	1.0	Ethinyloestradiol	35

INDICATIONS FOR USE

Contraception

The pill is indicated where maximum protection from pregnancy is required or where the woman wishes to use a method independent of intercourse.

It is particularly valuable, and is associated with lowest risk and fewest adverse side effects, in the young sexually active healthy non-smoking woman who is sufficiently motivated to be a reliable pill taker.

Gynaecological conditions

The lowest dose preparations are generally not very effective in the treatment of gynaecological conditions. It may be good practice to prescribe higher dose pills or even give the pill to women with one or more relative contraindications where it is indicated for 'treatment' and not solely for contraception.

OCs can be used to:

1. Relieve dysmenorrhoea.
2. Control menorrhagia. If fibroids are present and surgery is not indicated choose a relatively progestogen-dominant pill (pp. 51, 52, 70).
3. Relieve premenstrual tension.
4. 'Regulate' menstrual cycles. Periods become regular during therapy but do not necessarily continue to be so after the pill is stopped. Use of the pill *solely* to correct cycle pattern is not advisable. The reason for cycle irregularity or secondary amenorrhoea should be determined before the pill is prescribed.
5. Control endometriosis (progestogen-dominant brand).
6. Control functional ovarian cysts.

A specialist's advice should be sought when appropriate particularly for numbers 2, 5 and 6.

CONTRAINDICATIONS

All those involved in family planning should be aware of the content of the manufacturers' leaflets which are included in every packet of pills, and should be prepared to discuss them with the patient.

Absolute contraindications

1. Past or present circulatory disease.
   a. Arterial or venous thrombosis and any disease predisposing to thrombosis, e.g. the crises of sickle-cell disease. (Sickle cell trait, however, is *not* a contraindication to OC use)
   b. Crescendo migraines; any migraine requiring an ergotamine-containing drug; migraine of the focal type.
   c. Transient ischaemic attacks unrelated to headaches.
   d. Severe or combined *risk factors* for arterial disease (Table 4.2).
   e. Valvular heart disease, especially with pulmonary hypertension.
2. Diseases of the liver whether pill-induced or not.
   a. History of cholestatic jaundice of pregnancy.
   b. Infectious hepatitis — until six months after liver function tests (LFTs) have returned to normal.
   c. Cirrhosis, porphyrias, tumours of the liver, congenital liver disease, chronic active hepatitis.
   d. Presence of abnormal LFTs.
3. Undiagnosed genital tract bleeding.
4. Recent hydatidiform mole — until the beta-human chorionic gonadotrophin (hCG) test is negative (p. 7).
5. Actual or possible pregnancy.
6. History of any serious condition affected by sex steroids and occurring or worsening in a previous pregnancy, e.g. herpes gestationis, chorea and otosclerosis. See also 2a above.
7. Important conditions or side effects related to previous OC use, e.g. raised blood pressure (BP), liver adenoma.
8. a history of breast cancer — see relative contraindications.

Several of the above contraindications, such as 3, 4, and 5 are not necessarily permanent.

Relative contraindications

1. Risk factors for arterial cardiovascular system (CVS) disease. The main risk factors are listed in Table 4.2. Possession of one of these factors constitutes a relative contraindication to the pill. If present to a marked degree it becomes absolute. Multiple-risk factors may

Table 4.2 Risk factors for arterial CVS disease

Nature of risk factor	Absolute contraindication	Relative contraindication	Remarks
Family history of arterial disease (heart attack or stroke) in a first-degree relative	Known atherogenic lipid profile	Normal blood lipid profile (or not tested and first attack was in family member over age 45)	If available arrange test for plasma lipids if first attack <45 in family member
Diabetes mellitus	Severe, or diabetic complications present (e.g. retinopathy, renal damage)	Not severe/labile, and no complications	POP* usually better choice
Hypertension	Diastolic BP ⩾ 95 mmHg	Diastolic BP < 95 mmHg (but see p. 90 for implications of BP rise *de novo* during OC use)	POP* better choice
Cigarette smoking	⩾ 50 cigarettes/day	< 50 cigarettes/day	
Increasing age	⩾ 45, nonsmokers	35–45, nonsmokers	Smokers should avoid/ discontinue OCs *10 years earlier*
Excessive weight	⩾ 50% above ideal for height	< 50% above ideal	Extra caution if varicose veins *also* present

*Progestogen only pills.

contraindicate the use of the pill for contraception. However blood group O appears to be protective against both arterial and venous disease. Therefore a woman with this blood group might be allowed to continue using the pill despite risk factors which would normally contraindicate its use.

2. Age/smoking and duration of OC use. See page 93 (Table 4.6) for suggested scheme of management.

3. Varicose veins. The pill is not contraindicated in women with varicose veins unless complicated by attacks of phlebitis. Veins can become more prominent on the pill and may be a marker of other factors associated with venostasis especially obesity.

4. Long-term immobilisation, see page 92.

5. Before and after major surgery and treatment of varicose veins (p. 91).

6. Hyperprolactinaemia: only if the specialist agrees.

7. Sex steroid-dependent cancer, e.g. thyroid, ovary or breast. *Specialist advice* should be sought before prescribing OCs. Many believe that a history of breast cancer is an absolute contraindication to OC use. Breast biopsy showing epithelial atypia is considered an absolute contraindication by some (Kalache et al 1983).

8. Oligo/amenorrhoea — this should be investigated (p. 94) but the pill may be prescribed subsequently.

9. Severe depression (p. 66).

10. Chronic systemic disease (pp. 74, 95, 96). Evidence is emerging that Crohn's disease is a strong relative contraindication to the pill.

In some conditions such as diabetes mellitus and renal disease pregnancy is often contraindicated and may even be dangerous. The likelihood of adverse effects from the pill has therefore to be balanced against its benefits, especially its reliability as a contraceptive.

Since there is no evidence that the pill adversely affects the prognosis in patients with conditions such as multiple sclerosis and Hodgkin's disease it may be prescribed. Careful supervision and regular reassessment are important.

11. Diseases requiring long-term treatment with drugs which might interact with the pill (Table 4.4).

12. Conditions which impair absorption of oral contraceptives, e.g. some operations for obesity.

13. Some authorities suggest that the pill should not be prescribed a) if a first degree relative has had breast cancer, b) during monitoring of abnormal cervical smears, c) following definitive treatment of cervical intra-epithelial neoplasia. These views seem unnecessarily restrictive

in the light of our present knowledge (pp. 70, 71, 72), but clearly such women need special monitoring.

14. The woman's own uncertainty about the safety of the pill.

ADVANTAGES

1. Reliable, reversible, convenient, non-intercourse-related method.

2. Periods become more regular. Blood loss is reduced, decreasing the incidence of iron deficiency anaemia.

3. Menstrual and premenstrual symptoms such as dysmenorrhoea and premenstrual tension are often relieved.

4. Ovulation pain is abolished.

5. Unlike most other commonly used drugs there is no acute toxicity as a result of overdose, except vomiting and withdrawal bleeding in prepubertal girls.

6. Decreased incidence of:

    a. benign breast disease

    b. functional ovarian cysts

    c. pelvic inflammatory disease

    d. ectopic pregnancy since ovulation is inhibited and also as a long-term consequence of c.

    e. seborrhoeic conditions including acne (only with oestrogen-dominant pills)

7. Protection against carcinoma of the ovary and endometrium. This has been shown in at least 9 studies relating to the ovary and 7 to the endometrium. A duration-of-use effect providing a 2–3-fold reduction in the risk of both conditions after 5 years and an ex-use effect persisting for at least 10 years have been demonstrated.

8. Other possible benefits have been identified but have yet to be confirmed — protection against rheumatoid arthritis, thyroid disease, duodenal ulcer, endometriosis, trichomonal vaginitis and toxic shock syndrome.

DISADVANTAGES RELATED TO SYSTEMIC EFFECTS

It is not surprising that a combination of steroids which proves so effective in controlling reproduction also affects other physiological systems. Epidemiological evidence of adverse effects relates mainly to pills containing 50 μg or more of oestrogen. Well over 90% of women now use pills containing only 30 or 35 μg of oestrogen.

There has also been a welcome trend towards reducing the dose/potency of progestogen. High doses of progestogen tend to raise plasma insulin levels and lower the level of high density lipoprotein cholesterol (HDL-C), metabolic changes which are both believed to favour atherosclerosis. Epidemiological evidence on the incidence of hypertension and arterial disease shows a confirmatory association with increasing progestogen dose.

Lowering the dose of both components leads to a smaller alteration in many biochemical variables, including those just mentioned. It is believed that the use of such low-dose formulations for the great majority of women will reduce still further the low rate of serious adverse effects already documented for 'stronger' pills.

Clinical experience also suggests that these lowest dose preparations are less likely to cause the so-called minor side effects, such as nausea and weight gain.

Metabolic effects

These are numerous (Table 4.3). Many are similar to those found in normal pregnancy. Although this may be reassuring, it should be remembered that pregnancy has its own hazards often not unlike those of the pill.

The Liver

Contraceptive steroids are metabolised by the liver and affect its function. The resultant effects on the metabolism of carbohydrates, lipids, plasma proteins, amino acids, vitamins, enzymes and the factors concerned with coagulation and fibrinolysis explain the majority of the adverse effects, particularly on the CVS.

Increase in appetite and weight gain, with or without fluid retention, which occur in a proportion of OC users are also based on metabolic changes which are as yet not fully defined.

When specimens are sent to the laboratory it is important to state if the woman is taking the pill (Table 4.3).

Cardiovascular system

The incidence of the following conditions is increased in users of the pill.

Venous disease

Deep venous thrombosis; pulmonary embolism; thrombosis in other veins, e.g. mesenteric, hepatic or retinal.

Table 4.3 Some metabolic effects of combined oral contraceptives

	Blood level	Remarks
Liver		
Liver functioning		These many changes cause no apparent long-term damage to the liver itself. The liver is involved, however, in the production of most of the changes in blood levels, of substances shown elsewhere in this table, including the important changes in carbohydrate and lipid metabolism, and coagulation factors
a) generally	Altered in all users	
b) specifically		
Albumen	↓	
Transaminases	↑	
Amino acids	Altered	
Homocysteine	↑	
Blood glucose after carbohydrate ingestion	↑	
Blood lipids	Altered, mostly ↑	These changes, barely detectable with the latest pills, may partly explain the increased risk of arterial disease with higher dose preparations
HDL-cholesterol — in low oestrogen/progestogen-dominant OCs	↓	
Clotting factors		
a. generally	mostly ↑	Both the pill and smoking affect these interrelated systems. Fibrinolysis is enhanced in the blood, but reduced in the vessel walls
b. specifically		
Antithrombin III (anticlotting factor)	↓	
Fibrinolysis	↑	
Tendency for platelet aggregation	↑	
Hormones		
Insulin	↑	These hormone changes are related to those affecting blood sugar and blood lipids (above)
Growth hormone	↑	
Adrenal steroids	↑	
Thyroid hormones	↑	
Prolactin	↑	
Luteinising hormone (LH)	↓	These effects are integral to contraceptive actions. However, the first three tend to rise in some women during the pill-free week. Hence any effective lengthening of the pill-free time may lead to an LH surge and ovulation (Guillebaud 1984 p. 51)
Follicle-stimulating hormone (FSH)	↓	
Endogenous oestrogen	↓	
Endogenous progesterone	↓	

Table 4.3 (Contd.)

	Blood level	Remarks
Minerals and vitamins		
Iron	↑	This is a good effect for women prone to iron-deficiency
Copper	↑	
Zinc	↓	Effects unknown, but not believed to cause any health risk for most pill-users. Pyridoxine is discussed on page 87
Vitamins A, K	↑↓	
Riboflavine, folic acid	↓	
Vitamin B$_6$ (pyridoxine)	↓	
Vitamin B$_{12}$ (cyanocobalamin)	↓	
Vitamin C (ascorbic acid)	↓	
Binding globulins	↑	These globulins carry hormones and minerals in the blood. Because their levels increase in parallel with the latter, the effective blood levels of the hormones or minerals are usually not much altered.
Blood viscosity	↑	This retention of fluid explains some of the weight gain blamed on the pill (pp. 62, 68)
Body water	↑	
Factors affecting blood pressure		
Renin substrate	Altered	Changes do not correlate as well as expected with the incidence of frank hypertension (p. 65)
Renin activity	↑	
Angiotensin II		
Cardiac output		
Immunity/allergy		
Number of leucocytes	↑	See page 74
Immunoglobulins	Altered	
Function of lymphocytes	Altered	

Notes: 1. In the table ↑ means the level usually goes up, ↓ down.
2. 'Altered' means that the changes are known to be more complex, with both increases and decreases occurring within the system.
3. The changes are generally a) within the normal range, b) similar to those of normal pregnancy

The metabolic basis for these problems is an oestrogen-induced alteration of clotting factor levels tending to promote coagulation. One of the most important changes is a reduction in the level of antithrombin III. Platelet function is also modified. There is evidence of compensatory increased fibrinolytic activity, which may explain the rarity of overt disease in most pill users.

The risks of venous thrombo-embolism are enhanced by tissue damage and immobility (pp. 91, 92).

Arterial disease

This includes myocardial infarction; thrombotic stroke; haemorrhagic stroke including subarachnoid haemorrhage although the data here are less convincing; other arterial disease such as thrombosis of e.g. mesenteric or retinal arteries, and Raynaud's disease. It appears that the predisposition to arterial thrombosis results both from the metabolic changes (principally progestogen-related) which may promote atherosclerosis and the above oestrogenic changes affecting blood coagulability, the latter leading to thrombosis superimposed on the damaged vessel wall. Both the oestrogen and progestogen content are therefore involved, hence the modern teaching to use the lowest possible dose of each (p. 75).

Hypertension

In most women on the pill there is a slight measurable increase in both systolic and diastolic blood pressure within the normotensive range. Approximately 2.5–5.0% become clinically hypertensive the rate increasing with duration of use. The measurable metabolic changes in the renin-angiotensin system of the former are not obviously different from the hypertensive group.

Predisposing factors for OC-induced hypertension include a strong family history, and tendency to water retention and obesity.

Toxaemia of pregnancy *does not* predispose to hypertension during OC use (Pritchard & Pritchard 1977).

Hypertension is an important risk factor for heart disease and for both types of stroke.

Relative risk of cardiovascular disease

Epidemiological estimates vary but the overall mortality risk ratio from the prospective study of the Royal College of General Practitioners (RCGP 1981) was 4.0 for all vascular disease with risk ratios for the individual conditions varying between 1.5 and 6.0. There were very few deaths or incidents of disease under the age of 35. Smoking not only increased the attack rate for arterial diseases but also the case-fatality rate (RCGP 1983).

See Table 4.2 for risk factors.

Relative risks among current users. With the possible exception of strokes (RCGP 1983) the relative risk of CVS disease does not appear to increase with increasing duration of pill use provided those who develop hypertension are identified and discontinue OCs. The important factor is the age of the woman, not the number of years that she has been on the pill.

Relative risks among ex-users. There is no 'carry over' effect for venous thrombo-embolism beyond one to two months. There is such an effect for cerebrovascular disease only, with a lower risk ratio than for current users, but persisting for up to six years after stopping the pill (RCGP 1983). A similar phenomenon was shown in a case control study for myocardial infarction (Slone et al 1981), suggesting the (as yet unconfirmed) possibility that this effect was greater the longer the pill had *previously* been used. The data in both studies relate mainly to smokers above the age of 35 taking pills with higher doses of both oestrogen and progestogen than are now advised. It is believed that current low doses will reduce any ex-use effect.

Central nervous system

Depression

This was a relatively common complaint among women on high-dose pills but is less frequently reported by those using low-dose preparations. Although depression is commoner among women on the pill than among non takers, the causal factors may be the general life-circumstances of the pill-taker rather than the pill itself.

Depression may be related to altered tryptophan metabolism and hence is sometimes relieved by pyridoxine (p. 87).

Paradoxically some very depressed women find the pill relieves them of one of their great fears — that of unwanted pregnancy — and they therefore find it a very acceptable form of contraception.

Changed libido

Loss of libido is occasionally reported, particularly among those who are also depressed. This may be explained by psychological problems associated with pill taking such as 'maximum reliability means no gamble — no fun', by anxiety about its dangers magnified by the media, fears about subsequent fertility, or by guilt about using contraception at all.

For other women libido may be increased because the method is so reliable, requires no action related to sexual activity, and often reduces premenstrual tension.

Headaches (including migraine)

After depression, this was the second commonest reason for women stopping OCs in the RCGP study. Headaches often occur during the seven pill-free days (p. 87). Careful assessment is essential (p. 74). Reports by Bickerstaff (1975) and others suggest that the following may give advance warning of an impending stroke in pill takers:

1. First-ever occurrence of a severe migraine.
2. Crescendo migraine.
3. In a migraine sufferer, a change in the character of associated symptoms from *diffuse* to *focal*. These are symptoms suggesting transient cerebral ischaemia and may include unilateral loss of sensation, severe paraesthesiae or weakness, loss of a field of vision, dysphasia, focal epilepsy, and loss of consciousness.

Unilateral headache is common in migraine and does not have this significance.

4. Any similar symptoms suggestive of a transient ischaemic attack even in the absence of headache.
5. Any migraine severe enough to require ergotamine therapy.

Such symptoms have obvious implications for management (pp. 58, 87, 90) but if the woman can live with any other variety of headache, the pill can normally be started or continued.

Epilepsy

This condition is not initiated by OCs. In someone who has epilepsy the attack rate is often reduced though rarely it may be increased. Antiepileptic therapy is one of the few indications for a relatively high-dose pill (p. 82).

Chorea and benign intracranial pressure

The incidence of these rare conditions is increased in users of the pill but both improve if OCs are discontinued.

Eyes

Minimal water retention can lead to slight corneal oedema, and result in discomfort or corneal damage in those who wear contact lenses. With modern soft lenses and low-dose pills this problem is now rare.

The catastrophies of retinal artery or vein thrombosis and bleeding are even rarer but may be related to OC use.

Respiratory system

Allergic rhinitis and asthma

There is some tenuous evidence of a causal association between OC use and these conditions particularly the former.

Gastro-intestinal system

Nausea and vomiting

Nausea may occur in the first cycle and occasionally recur with the first few pills of each packet. Vomiting is most unusual. Both symptoms are related to the oestrogen component of the pill and are rare with low dose preparations (p. 88). Severe vomiting may interfere with OC absorption and lead to breakthrough bleeding or spotting (p. 86).

Weight gain

Increase in weight is unusual with modern pills although it is frequently and unjustifiably expected. It is commonly due to overeating, sometimes associated with an increase in appetite on starting the pill.

Cholestatic jaundice

This is commoner among OC users and also in pregnancy. A past history in either context contraindicates the pill.

Gall stones

Latest reports imply that the increased risk of gall-stones among OC users is significant only during the early years of pill taking. This suggests that the risk applies only to predisposed women.

Liver tumours

The relative risk of benign adenoma or hamartoma is increased by OC use. However the background incidence is so small (1–5 per 1 000 000 women per year) that the OC-attributable risk is minimal. Most reported cases have been in long-term users of relatively high-dose pills.

Cases most commonly present with an acute abdomen due to intraperitoneal bleeding. The liver is always enlarged, often with a hepatic bruit. Liver function tests are disturbed.

There is no proof of a causative link between primary carcinoma of the liver (an even rarer condition) and OC use although there have been some suggestive reports.

Urinary system

Several studies show that urinary tract infections are more common in OC users than in controls. Although women on the pill may have more frequent intercourse, predisposing them to 'honeymoon' cystitis, evidence of an increased incidence of symptomless bacteriuria in OC users suggests a causal link.

Genital system

Menstrual cycle

Most symptoms associated with the menstrual cycle are improved (p. 61). Only a minority of pill takers suffer from a type of premenstrual syndrome or dysmenorrhoea. Postpill amenorrhoea, breakthrough bleeding and absent withdrawal bleeding are considered on pages 93 and 86.

Vaginal discharge

Low-dose pills usually do not increase or decrease vaginal discharge. However, some women on progestogen-dominant pills may complain of vaginal dryness while those on oestrogen-dominant preparations may develop cervical erosion with resultant profuse clear mucoid discharge.

It is generally believed that moniliasis is commoner among OC users but the evidence is conflicting.

OCs may provide some as yet unexplained protection against trichomonal vaginitis (p. 61) but not against other sexually transmitted diseases. The possibility that the discharge may be due to infection with gardnerella, chlamydia or the gonococcus must always be borne in mind.

OCs do however protect the *upper* genital tract from pelvic infection.

Fibroids

Oestrogen-dominant OCs may cause them to increase in size. Progestogen-dominant brands are thought to reduce their rate of growth and associated symptoms.

Since there is no evidence that low-dose pills affect their growth either way patients with small fibroids may therefore be given the pill.

Malignant disease

There is no *definite* evidence that taking the pill increases the risk of cancer of any kind.

Carcinoma of the ovary and the endometrium. (p. 61)

Carcinoma of the cervix. The literature is complex and often contradictory but some studies suggest a possible association between this cancer and its premalignant state and oral contraceptive therapy. Vessey et al (1983) showed an effect of duration of pill use in a prospective study involving about 10 000 women. In this study all 13 cases of invasive cancer occurred in the OC group. But there is no certainty yet that the confounding variables — especially young age at first intercourse and multiple sexual partners — were not more frequent in the OC users than in the IUD users, with whom they were compared. Moreover these factors have a far greater effect on the risk than the 2–3-fold increase after 8 years demonstrated in Vessey's study.

The main implications for prescribers are:

1. A cervical smear should be taken when the pill is initially prescribed or at a subsequent visit in the near future. A second smear within one year minimises the problem of false negative reports.

2. Smears should be repeated at least every five years thereafter (Ch. 15).

3. Women at high risk, eg admitted multiple partners or infected with herpes or wart virus, require more frequent screening.

4. It is acceptable practice to continue OC use during the careful monitoring of cervical smear abnormality or subsequent to the definitive treatment of cervical intraepithelial neoplasia. However,

some consider these to be *relative* contraindications to the pill (pp. 60, 61).

5. The evidence does not justify advising women to discontinue pill use on these grounds after a specific number of years.

Choriocarcinoma. Following the diagnosis of trophoblastic disease if OCs are prescribed before hCG levels are undetectable, the risk of choriocarcinoma is doubled (Stone et al 1976) (p. 58).

Breasts

Although benign breast disease may be controlled by OCs, some women complain of breast tenderness and swelling. These symptoms can occur with any pill formulation but seem to be particularly associated with the last phase of triphasic and biphasic brands.

The pill reduces the volume and quality of milk in many lactating women, particularly when lactation is not well established.

Galactorrhoea among pill takers is rare and needs investigation.

Carcinoma of the breast

The literature is copious, complex and contradictory. In assessing a possible link between the pill and breast cancer many factors have to be taken into account:

1. The long latent period between exposure or repeated exposure and the disease.

2. That different effects are possible according to:
   a. The time of life when exposure occurred — in this context, particularly before full-term pregnancy or possibly before the age of 25.
   b. Other risk factors, especially a family history of breast cancer, or personal history of benign breast disease.
   c. Change in pill formulation — both the oestrogen and progestogen content of pills have changed over the years and different preparations were popular in different countries.

3. Confounding factors — the pill may act as a 'marker' for some other risk factor, known or unknown.

Until recently it was generally agreed that there was no significant aetiological link between OCs and breast cancer, though caveats were expressed about exposure in the young (Kalache et al 1983). In an article in *The Lancet* in October 1983 Pike et al suggested that young women who took the pill for prolonged periods before the age of 25 were at increased risk of developing cancer of the breast. His study has been criticised on many counts, chiefly that the interviews were conducted by telephone; only 314 case-control pairs were constructed although 510 eligible cases were originally identified; and all the cases used had to be alive and reasonably well. If reports suggesting that OCs have a beneficial effect on tumour growth and spread prove to be accurate (Kalache et al 1983), this fact could seriously bias results by inflating the number of pill-users available for study among the cases.

Pike's findings have been substantially supported by McPherson et al (1983) both studies showing a duration-of-use effect.

However neither the major 'cancer and sex hormones' project (CASH) of the Centers for Disease Control in Atlanta, USA, nor the Boston Drug Epidemiology Unit is able to show any excess risk of breast cancer, whether the pill was used at a young age or before the first full-term pregnancy (Lincoln 1984).

An attempt by Pike to identify brands of pills particularly likely to be associated with breast cancer risk, on the basis of their progestogen potency, produced great confusion. He used the Greenblatt 'delay of menses test' to measure potency but as that test is now largely discredited the resultant recommendations are not relevant.

Current interpretation of Pike's paper in context with other published work suggests that *if* OCs do in fact increase breast cancer risk (in the young, or indeed in women of any age) this risk will be minimised by the use of pills which contain the lowest available dose of both progestogen and oestrogen. It is possible that these low dose pills may even protect against breast cancer. Fortunately these are precisely the same pills which are recommended for use to reduce the risk of CVS disease (p. 65)

It may now be preferable:

1. To prescribe the very *lowest dose* available of any chosen *progestogen* combined with 30–35 µg of oestrogen for most women starting the pill for the first time.

2. Ideally to prescribe pills for a limited time to women with a family history of a first-degree relative with breast cancer.

It is acceptable practice, having regard to the pregnancy risk and problems associated with alternatives, to prescribe lowest dose pills to the young, including teenagers, without any arbitrary time limit.

All women should be taught breast self-examination.

Musculoskeletal system

Oestrogen given in high doses to young female animals can lead to premature closure of the epiphyses. There is no evidence that a daily dose of 30 or even 50 μg of ethinyloestradiol has this effect in postpubertal girls. Their growth will not be stunted if they take the pill. Carpal-tunnel syndrome is more frequent in pill takers. There is evidence that the pill may protect against rheumatoid arthritis (pp. 61, 74). Leg pains and cramps — see page 89.

Cutaneous system

Chloasma/melasma

'Pregnancy mask' may develop in women on the pill after excessive exposure to sunlight, irrespective of whether or not it occurred in a previous pregnancy. It appears that both oestrogen and progestogen can be contributory. The condition may be slow to fade after the pill is stopped.

Photosensitivity

This may rarely be the first manifestation of one of the porphyrias.

Acne/greasy skin, hirsutism

These androgenic effects are associated only with oestrogen-deficient/progestogen-dominant pills.

Malignant melanoma

Increased risk of malignant melanoma has been suggested in OC users. There appears to be a 2–3 fold risk of the superficial spreading type. It is not clear whether this is related to the pill or associated factors such as exposure to bright sunlight.

Other skin conditions

Telangiectasia, rosacea, eczema, neurodermatitis, erythema nodosum, erythema multiforme and herpes gestationis may all be causally associated or exacerbated by oral contraceptives.

Infections and inflammations

Some studies suggest that OC users are more likely to suffer from infections such as chickenpox, gastric 'flu, respiratory and urinary tract infections, and also inflammations such as tenosynovitis and a form of allergic polyarthritis.

There is also evidence that Crohn's disease may be more frequent in pill takers. This is usually the non-granulomatous type and resolves if OCs are stopped (Rhodes et al 1984).

These and other effects, including the apparent beneficial effect on rheumatoid arthritis and thyroid disease, are believed to indicate that the pill can modify immune mechanisms.

CLINICAL MANAGEMENT

Assessment

Suitability for the pill is based on history and physical examination.

History

General. Take a full history (Ch. 2). Pay particular attention to those conditions which might contraindicate OC use and the importance for this individual of avoiding pregnancy.

Specific. The following factors and conditions may influence the initial decision to start the pill. They also form the baseline on which the significance of side-effects can be evaluated in the future and against which a decision to change the pill or the method can be made.

1. Social factors.

2. Family history, especially of CVS disease including hypertension, and of breast cancer.

3. Aspects of fertility: recent delivery or termination of pregnancy.

4. Current ailments and investigations.

5. Current drug therapy.

6. Allergies: hay fever, skin troubles, asthma.

7. Current or past CVS disease.

8. Headache/migraine: frequency, site, timing, severity, relation to menstrual cycle, initiating factors, therapy, presence of focal symptoms in the past (p. 67).

9. Legs: pain and/or tenderness, swelling, varicose veins, phlebitis.

10. Jaundice: age at which it occurred, nature of illness (infective or obstructive).

11. Epilepsy: age of onset, type of disease, degree of control, treatment if any.

12. Use of contact lenses: date of last check, any symptoms.

Examination

1. Observation on general health.
2. Record weight and blood pressure.
3. Although it may be reassuring to establish that the pelvis is normal, many women dislike 'internal' examinations. Pelvic examination should not be insisted upon at the first visit provided the patient gives a normal menstrual history with a normal last menstrual period, indicating that she is not pregnant. Pelvic examination may then be delayed until a later date.
4. Carry out a cervical smear according to the clinic or practice policy for a woman in that age group.
5. Breast examination is recommended. Women should be taught breast self-examination (Ch. 15).
6. Further investigation where appropriate, e.g. liver function tests for those with recent jaundice, glucose tolerance test for those who are grossly obese or with a history suggesting latent diabetes.

Choice of pill

A few patients are only suited by certain formulations and some women are suited by none. Much has been written about matching pills to particular hormonal profiles, but the systems have no practical value for the initial selection of the low-dose pills now in use.

Each doctor needs to be familiar with the composition of the available preparations. Women may react unpredictably and several types may have to be tried before a suitable one is found.

It is wise to identify the lowest effective dose for each woman. This will minimise adverse side-effects, both serious and 'minor', and normally means starting with the lowest dose available:

1. The dose of oestrogen should be less than 50 μg, except in special circumstances such as
    a. where there is long-term use of a drug with which the pill may interact (Table 4.4),
    b. where there is any reason to suspect malabsorption,
    c. when a 30 μg preparation cannot control the cycle after at least three month's trial.
2. The lowest dose of a given progestogen within each group of low oestrogen brands should be used (Table 4.1).

If this policy is adopted almost the only reason for moving to a higher dose pill is poor cycle control which the woman cannot tolerate.

This policy requires to be flexibly applied. All women should be advised about probable spotting and breakthrough bleeding or absent withdrawal bleeds at least in early cycles. Those considered to be unlikely to tolerate, or become confused by, such symptoms and all who are unlikely to be consistent pill takers should be given a higher dose (chosen from among the pills listed in Table 4.1) from the outset.

3. The progestogen-dominant brands Conova 30, Eugynon 30 and 50, Ovran 30 and Ovran are best avoided (National Association of Family Planning Doctors 1984).

4. Phased pills have a reduced margin for error if pills in the first phase are omitted. There is also an increased likelihood of errors in pill taking.

5. Psychology in pill prescription may be as important as physiology. Thus, complying with the woman's own personal choice, because a pill suited a friend or relative for example may save a lot of trouble.

6. Women who find pill taking difficult, especially those who are poorly motivated, may be helped by the everyday (ED) type of pill.

First choice of pill

To identify those women who are well suited by the lowest dose pills available and who hence run the lowest possible risk of long-term side effects, in my view one of the following pills should be the doctor's preferred initial choice but *not* for every case. See 2 above.

1. Brevinor or Ovysmen.
2. Logynon or Trinordiol.
3. Marvelon.

Loestrin 20 is rarely to be recommended because of its poor reputation for cycle control and some doubts about its efficacy. Loestrin 30, which contains a higher dose of both oestrogen and progestogen than Loestrin 20 is now available.

Administration

Regulations

Regulations concerning pill prescription are controlled by the Medicines Act 1968.

1. Only a medical practitioner may prescribe the pill.

2. A nurse may supply and resupply pills to the patient in accordance with the doctor's directions written on the prescription.

Oral contraceptives are Schedule 4B poisons in accordance with the Pharmacy and Poisons Act 1963. This requires that:

1. Every woman is examined before the pill is prescribed.

2. Prescriptions are in writing and must be dated.

3. Prescriptions are not indefinite in terms of quantity, duration and validity.

4. Each container should be labelled with the name of the patient, the product, the place and the date of issue.

In accordance with the Poison Rules of 1968, arrangements should be made for inspection of storage facilities at intervals of not more than three months by the pharmacist or person in control of the clinic. A written record of the inspections should be kept dated and signed confirming that the storage system is satisfactory.

Instructions to patients

Although pills are now dispensed in bubble packs, which should make it easier to take them properly, careful detailed teaching and explanation are still essential. Leaflets included in each packet are in small print and it is wise to explain them to each woman. This can be reinforced by giving her the appropriate method instruction leaflet produced by Family Planning Information Services (FPIS).

The need for regular pill taking at a time of day when it is easy to remember such as when cleaning one's teeth in the morning or evening must be stressed. Irregular pill taking leads to poor cycle control and unreliability.

Starting the pill

1. Monophasic pills:

    a. The first pill is taken on the first day of the next menstrual period. Manufacturers' literature suggests that starting should be delayed until the fifth day for some brands.

    b. The pill labelled for the appropriate day of the week is selected.

    c. One pill is taken daily for 21 (or 22) days at approximately the same time.

    d. After the packet is finished no pills are taken for 7 (or 6) days, after which a new packet is started.

    e. Everyday (ED) varieties contain 7 placebo tablets which are taken during the 7 days that would otherwise be 'tablet free'. With this regimen the next packet of pills is started immediately after the last one is finished.

    f. Contraceptive protection is immediate if the pill is started on the first day of a period. If starting is delayed until the fifth day, additional contraceptive precautions are needed for the first 14 days of the first packet. Extra precautions are also required for the first 14 days with ED varieties.

2. Triphasic pills: each packet contains 21 tablets of three different strengths identified by different colours (p. 56). It is most important that the pills are taken in the correct order.

a. The first pill, labelled '1', is taken on the first day of the period.

b. With Trinordiol and Logynon the appropriate day in the row of 'reminder blisters' is pricked as an indicator of the week-day on which pills number 8 and 15 should be taken and of the day on which each subsequent new packet is started. Pills are taken in the same way as monophasic pills starting with 6 brown tablets followed by 5 white tablets and finally 10 yellow tablets.

c. Packets of Trinovum are divided into 3 'snap-off' sections numbered 1, 2 and 3, each containing 7 pills which are either white or a shade of peach. The day of the week is marked against each tablet. The white tablets are taken first, followed by the 7 tablets from section 2 and then the 7 from section 3.

3. Biphasic pills (Binovum):

a. Each packet contains 21 tablets (7 white and 14 peach coloured) labelled with the correct day of the week.

b. The first white tablet is taken on the first day of the menstrual period followed by one white tablet daily for the next 6 days.

c. Once all the 7 white tablets are taken the first peach coloured tablet is taken from the second section and repeated daily till the packet is finished.

Starting after delivery

Ovulation has not been reported earlier than the 5th week postpartum. Because the risks of thromboembolism are greater in the immediate postpartum period, it is therefore preferable to avoid the combined pill initially. The fourth week after delivery is a good time to start it.

During lactation the progestogen-only pill (POP) is preferable (p. 101) — the combined pill adds nothing to the efficacy of the POP combined with full lactation especially if menstruation has not been re-established.

Starting after termination of pregnancy or miscarriage

The pill should be started immediately. It does not interfere with recovery, nor increase morbidity.

Changing preparations

When changing from one combined pill to another, one of the following regimens may be used. The first pill of the new packet is taken:

1. After the seven tablet-free days. Additional contraceptive precautions are required if the new pill has a lower dose of either oestrogen or progestogen than the previous preparation. Otherwise no additional extra precautions are required. _Or_

2. On the day immediately after the old packet is completed. No extra precautions are necessary. _Or_

3. On the first day of withdrawal bleeding following completion of the last pack. No additional precautions are required. The risk with this regimen is that if the woman does not get a withdrawal bleed that month she may delay too long before starting the new brand. She should be told not to wait longer than seven days before starting a new packet and then to take additional contraceptive precautions if the new pill is a low dose preparation.

When changing from a POP the first combined pill is taken on the first day of the next period or, if the woman is amenorrhoeic, immediately after the last pill of the POP packet.

Manipulation of the menstrual cycle

1. To postpone a period: The pill may be used to postpone menstrual bleeding for the social convenience of the woman, e.g. while she is on her honeymoon, taking part in sporting activities, sitting exams etc. She takes two packets of fixed dose pills consecutively with no break.

Phased preparations cannot be used in this way because if the woman takes two packets of these pills one after the other she is liable to develop breakthrough bleeding, since the first pills from the second packet contain a lower dose of progestogen than those she has just finished. Progestogen levels in the blood fall and withdrawal bleeding occurs — the very thing she wanted to avoid. Women on phased preparations may delay menstruation in one of two ways (Fig. 4.1):

a. Continue to take further tablets from the *final* phase of a new packet of the same brand of phased preparation.

b. Interpose between *two* packets of phased pills a packet of the fixed-dose brand most similar to the relevant *final* phase. This is Microgynon or Ovranette for Logynon and Trinordiol: Norimin or Neocon for Binovum and Trinovum. Contraceptive efficacy will be maintained but breakthrough bleeding may occur after the middle packet in this 'sandwich'.

2. Tricycle regimen: Some women who have severe menstrual or premenstrual symptoms also welcome the opportunity of menstruating less frequently. Others wish to avoid as much as possible the inconvenience of even light menstrual loss while using the pill. A well-tried system involves taking 4 packets of monophasic pills consecutively followed by a 7-day break. This adds up to a 91-day pill

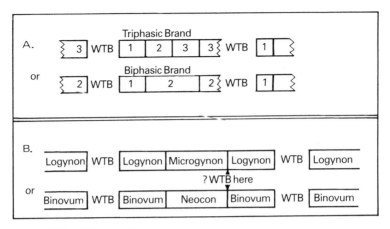

WTB = Withdrawal bleed

1, 2, 3 = Relevant phases of brand concerned

Fig. 4.1 Postponement of menstrual bleeding on a biphasic or triphasic pill

cycle and means that the woman has only 4 withdrawal bleeds a year. Since the regimen involves a larger annual intake of contraceptive steroid than with standard pill taking (which is contrary to the general principle) its use is usually restricted to:

a. Avoiding 'withdrawal' headaches and (nonfocal) migraines.

b. The treatment of endometriosis using a progestogen-dominant brand.

Triphasic or biphasic pills cannot be used for this regimen.

Management of missed pills

1. If a pill is missed for more than 12 hours after it was due to be taken contraceptive reliability is threatened and breakthrough bleeding (BTB) or spotting (Sp) may occur. The safest rule is as follows:

a. Take the overdue pill as soon as it is remembered.

b. Take the next pill on time and continue the rest of the pack.

c. Take extra precautions for the next 14 days, even if this includes some or all the 7 tablet-free days.

2. If several pills are missed in a row, 2 tablets should be taken together (these should be 2 of the most recent pills missed, hence the strongest, if a phased pill is in use), pill-taking resumed with the next pill that is due to be taken and the pack completed. Once again extra precautions should be used for the next 2 weeks.

3. If withdrawal bleeding subsequently fails to occur, check that the woman is not pregnant.

4. If the cause of 'missed pills' is lack of absorption due to vomiting or diarrhoea extra precautions should be used during the illness and for 14 days thereafter.

Drug interaction

The importance of potential reduction in the effectiveness of OCs resulting from the ingestion of other medications has come to light since the introduction of low oestrogen pills (Back & Breckenridge 1978, Back et al 1983). The interaction may occur at any stage between absorption and excretion and complex mechanisms are involved. Induction of liver enzymes and competition for binding sites play important roles.

Table 4.4 shows the more important drug interactions of this type. Those which are best established and most clinically relevant relate to rifampicin and anticonvulsant preparations.

A major problem is that women show enormous variability in the blood levels of both oestrogen and progestogen after ingestion of the same brand of pill. In addition, for any of the drugs in Table 4.4 individual women demonstrate different degrees of interaction which cannot be predicted clinically. In one study only 2 out of 13 women showed a lowering of the concentration of ethinyloestradiol while taking ampicillin 500 mg tds at the same time as the pill. This suggests that at most 1 in 7 women might be at risk of pregnancy. However, in practice it is wise to assume reduced protection and to advise all women to use extra precautions during treatment and for 14 days thereafter, even if this period extends into or beyond the 7 pill-free days.

Breakthrough bleeding may be a clinical marker of drug interaction and used as an indication to make appropriate alteration to pill prescription, or to advise a change of method.

OC steroids are also inhibitors of hepatic microsomal enzymes. They thus slightly lower the clearance of for example diazepam, prednisolone and probably other drugs. This may increase the risk of side effects and toxic manifestations. The effect is very unlikely to be noticed clinically.

As OCs tend to impair glucose tolerance, sometimes cause depression and raise the blood pressure, they naturally tend to inhibit the action of antidiabetic, antidepressant, and antihypertensive

Table 4.4 The more important drug interactions with oral contraceptives

Class of drug	Approved names of important examples	Main action	Clinical implications for OC-use
Drugs which may reduce OC efficacy			
Anticonvulsants	barbiturates (esp. phenobarbitone) phenytoin primidone carbamazepine ethosuximide	Induction of liver enzymes, increasing their ability to metabolise *both* OC steroids	50 μg oestrogen OCs can be used, increasing to 80 μg if BTB occurs. Sodium valproate is the only common anticonvulsant without this effect
Antibiotics			
a. Antitubercle	rifampicin	Marked induction of liver enzymes	Use alternative contraception
b. Broad-spectrum	ampicillin tetracyclines	Change in bowel flora, reducing enterohepatic recirculation of ethinyloestradiol (EE), only, after hydrolysis of its conjugates	Short courses —wisest to use additional contraception during illness + for 14 days Long-term low dose tetracycline for acne — no apparent problem, probably because resistant organisms develop, also 50 μg EE pill often used. Progestogen-only pill — believed to be unaffected by this type of interaction

Table 4.4 (Contd.)

Class of drug	Approved names of important examples	Main action	Clinical implications for OC-use
Hypnotics	chloral hydrate dichloralphenazone glutethimide	Induction of liver enzymes	Avoid these drugs in OC-users (alternatives available)
Miscellaneous Tranquillizers	All enzyme inducers chlorpromazine meprobamate benzodiazepines		Recommend alternative therapy *or* additional contraception during short-term treatment *or* use 50 μg oestrogen OC
Anti-inflammatory drugs Antifungal drugs Diuretics	phenylbutazone griseofulvin spironolactone		

Drugs which may increase OC efficacy

| | Ascorbic acid | Competition in bowel wall for conjugation to sulphate. Hence if vitamin C present, more EE available for absorption | Only applies to mega-doses (0.5–1 g daily)

Effectively results in the patient taking a high oestrogen OC. |
| | Co-trimoxazole | Inhibits EE metabolism | No effect on the progestogen.
None, if short course given to low-dose OC-user |

preparations. The first two types of interactions are generally only of minor importance, and it is unusual for an OC-user to be prescribed an antihypertensive drug (p. 91).

OCs also interact with ergotamine preparations (p. 67).

OCs impair to a variable extent the metabolism of warfarin and at the same time alter clotting factors. Since the interaction is completely unpredictable, this combination of drugs is best avoided.

Follow-Up

Patients should be seen three months after starting the pill (earlier if there are any relative contraindications) and six-monthly thereafter. At each visit it is important to assess the acceptability of the method and to check that it is being used correctly. Any newly-apparent risk factors must be noted and side effects such as rise in blood pressure, headaches, weight gain, dealt with as and when necessary.

Where no problems arise routine examinations should be kept to a minimum.

1. Routine weighing is unnecessary (Ch. 5).

2. Blood pressure should be recorded regularly during the first year and at least annually thereafter.

3. Screening tests should be carried out in accordance with standard procedures. (Ch. 15).

4. In the presence of symptoms, the appropriate examinations/investigations should be performed.

COMPLICATIONS AND THEIR MANAGEMENT

While women should be forewarned about potentially serious symptoms (p. 90) and feel free to report any side-effects which worry them it is important to avoid 'self-fulfilling prophecy'. It is helpful to explain to them that the low-dose pills we now prescribe not only minimise the risk of serious complications but also reduce the likelihood of practically all problems except those related to cycle control.

General Principles

In general terms if a change of pill seems appropriate to cope with a specific side effect try:

1. Changing to a pill with a lower dose of the same progestogen and/or oestrogen (if the woman is taking a pill containing 50 μg of oestrogen).

2. Changing to a different progestogen, again using the lowest available dose.

3. In certain specific instances choosing a pill according to its oestrogen or progestogen dominance depending on whether the symptom is believed to be related to relative oestrogen or progestogen excess (Table 4.5 and Ch. 3). However, this approach has doubtful scientific basis and strongly held clinical views are not backed up by good controlled studies using modern pills.

Table 4.5 Side-effects and their management

Symptoms	Conditions
Relative oestrogen excess	
Nausea	Benign breast disease
Dizziness	Endometriosis
Cyclical weight gain (fluid)	Fibroids
Bloating	
PMT & irritability	
Vaginal discharge	

Action
Change to progestogen-dominant OC such as: Norimin or Loestrin 20 or 30 (preferred), or Conova 30, Eugynon 30 (Ch. 3, Table 3.6)
(But: the latter will have relatively marked effects on HDL-C — Ch. 3)

Symptoms	Conditions
Relative progestogen excess	
Greasiness of skin/hair	Acne
Dryness of vagina	Hirsutism
Some cases of loss of libido	↓HDL-C
depression	
weight gain	
fatigue	

Action
Change to oestrogen-dominant OC such as: Ovysmen/Brevinor, Marvelon, Trinordiol/Logynon, Minovlar/Orlest (Ch. 3, Tables 3.5 and 3.6).

Using another drug to treat pill-induced side effects is nearly always bad practice. For example, if a woman complains of headaches try changing her pill or recommending the tricycle regimen rather than prescribing analgesics or antimigraine treatment.

If every brand of combined pill one tries causes side effects, the progestogen-only pill may prove satisfactory.

Breakthrough bleeding (BTB) and spotting (Sp)

Although these symptoms can signify that the blood level of OC steroids is too low to maintain the endometrium, this is an end-organ effect and does not necessarily correlate in a particular woman with an increased risk of pregnancy. Most women who become pregnant on the pill have had no previous BTB or Sp just as most patients with these problems do not become pregnant.

BTB and/or Sp are common during the first two or three cycles of pill-taking, especially in the first cycle with the 'day 1' start regimen. The woman should be forewarned, advised to continue regular pill-taking and not to stop in the middle of a packet. If Sp or BTB do occur she should be given a chart on which to record any further episodes of bleeding, and the situation reviewed after three months.

If these symptoms persist check the following points.
1. Is the patient taking her pills correctly?
2. Is there any reason to suspect malabsorption?
3. Could there be any drug interaction? (p. 81)
4. Check for gynaecological disease, especially neoplasia, by appropriate examination including cervical smear if indicated and referral to a gynaecologist if in any doubt.

There is no simple therapeutic formula for coping with unexplained BTB or Sp which may occur after many years of good cycle control on the pill.
1. If she is taking a monophasic pill try a phased variety.
2. If none of these preparations works:
    a. increase the progestogen component of the pill, or
    b. transfer to a similar brand containing a different progestogen, or
    c. use a pill with 50 μg oestrogen such as Minovlar or Gynovlar.
This is now one of the few indications for using such pills.

Absence of withdrawal bleeding

This is not dangerous, does not signify overdosage, nor is it related to postpill amenorrhoea. The incidence of amenorrhoea is very low with triphasic and biphasic pills.

Check that the woman is not pregnant, particularly if pills have been missed, possibly not absorbed, or where drug interactions may have occurred.

Ideally, if a patient misses a withdrawal bleed the next packet of pills should be delayed until the possibility of pregnancy can be excluded. This is often not practical and, if the pill is not started and if alternative contraception is not used, the danger of unplanned pregnancy is very real. However, if two withdrawal bleeds have been missed the patient should not start her next packet of pills until she is assured that she is not pregnant.

There is no magic formula for coping with patients who have persistent amenorrhoea on the pill. A triphasic or biphasic preparation may be tried. If neither works try increasing the dose first of progestogen and then if necessary of oestrogen.

Other side effects

Central nervous system

Headaches. Since headaches are possibly associated with increased risk of stroke and the aetiology of pill-induced headaches is not fully understood, any woman reporting headaches on the pill must be taken seriously. The headache pattern should be carefully assessed (p. 74) and compared with the pattern of headaches before starting the pill. It is often helpful to record on a chart when headaches occur.

If she reports severe migraine for the first time, or if focal symptoms suggestive of transient ischaemia occur, the pill should be stopped immediately and replaced by another method of contraception.

If headaches appear to be pill-induced try the effect of:

1. A different variety of monophasic pill or a phased preparation.
2. Changing to a progestogen-only pill.
3. Increasing the number of pills taken to 24 followed by 4 pill-free days.
4. The tricycle regimen.

Measures 3. and 4. are particularly effective in reducing the frequency of headaches which occur during pill-free days.

If the headaches are very severe change to a non-hormonal method.

Depression. Depression may be alleviated by:

1. Lowering the dose of progestogen
2. Pyridoxine (vit. B_6) 50 mg daily. This may take up to two months to be effective.

Loss of libido. Marital and family circumstances and psychosexual aspects of the relationship should always be fully discussed. If vaginal soreness and dryness are caused by thrush the infection should be treated. Lubrication with a bland jelly (KY) often helps. Changing to an oestrogen-dominant pill is frequently beneficial (p. 85, Table 4.5).

Eye changes

If any acute visual disturbance occurs, the pill should be stopped pending further investigation. 'Blurring' of vision would be of far less importance than, for example, loss of field of vision (p. 67).

If corneal irritation occurs in wearers of contact lenses prescribe a pill containing the lowest possible dose of both steroids. If symptoms persist the wearer has to decide whether to give up her contact lenses or the pill otherwise corneal ulceration and scarring may result.

Gastro-intestinal system

Nausea and vomiting. Nausea is less likely to occur if the pill is taken at night rather than in the morning. If persistent nausea occurs try changing to a pill with less or a different oestrogen first. (Ortho-Novin 1/50 and Norinyl-1 are the only pills with mestranol. The others all contain ethinyloestradiol.)

Vomiting starting for the first time after months or years of trouble-free pill taking should not be attributed to the pill.

Weight gain. The patient should be advised about her diet and if this does not produce the desired result the pill should be changed to one containing a lower dose of the same progestogen or to one with a different progestogen.

The progestogen-only pill may be tried in women who continue to put on weight on different brands of the combined pill.

Jaundice. Discontinue the pill immediately. If a diagnosis of infective hepatitis is made the pill should not be restarted until six months after liver function tests have returned to normal. A diagnosis of cholestatic jaundice means that the pill is contraindicated and an alternative method of contraception should be chosen.

Gall stones. It is reasonable for a woman who has had definitive treatment by surgery to take a low-dose pill provided the surgeon approves.

Crohn's Disease. Stop the pill.

Genital system

Vaginal discharge. It is important to establish the cause by pelvic examination, speculum examination of the cervix and microbiological examination when necessary.

Symptomatic cervical erosion may be treated with cryocautery as an outpatient.

If infection is present the appropriate treatment should be prescribed.

Where the diagnosis is in doubt or the condition does not respond to treatment the patient should be referred to a department of genito-urinary medicine. This is particularly important in the patient with trichomoniasis or genital warts as in both cases the partner may require treatment. In addition both conditions can act as markers for more serious sexually transmitted disease.

Fibroids (p. 70). Although there is no evidence that low-dose pills increase the rate of growth of fibroids it is wise to arrange six-monthly pelvic examinations by the same observer just to be sure.

Breasts

Sudden enlargement of the breasts may be the first sign of pregnancy.

Breasts which are tender on touch or pressure, whether they are enlarged or not, often respond to a pill with a lower dose of progestogen. Painful breasts are often relieved by increasing the dose of progestogen. A progestogen-dominant pill is indicated for women with a tendency to benign breast disease (Table 4.5).

Breast symptoms may continue no matter what preparation is prescribed and sometimes oral contraceptives have to be abandoned.

Galactorrhoea is rare but needs investigation.

Management of the patient in whom a discrete breast lump is found is dealt with in Chapter 15.

Musculoskeletal system

Leg pains and cramps. Careful examination and assessment of patients presenting with these symptoms is important.

If bilateral, note any altered physical activity, water retention, weight increase, varicose veins, chilblains or Raynaud's disease.

If unilateral, deep venous thrombosis must be excluded.

If in any doubt about the diagnosis the pill should be stopped and the patient referred for further investigation including venogram if possible. Her contraceptive future depends on an accurate diagnosis being made.

Cutaneous system

Chloasma. Mild degrees of chloasma can be masked by carefully applied cosmetics. If it is causing the woman distress a different pill can be tried but often no benefit is gained by changing from one low-dose preparation to another. A progestogen-only pill benefits some women.

Changing to a non-hormonal method is often the best plan but even then the pigmentation may be slow to fade.

Depigmenting creams and lotions should be avoided.

Acne/greasy skin and hirsutism. An oestrogen-dominant pill is indicated (Table 4.5). Good results have been reported using Logynon, Trinordiol and Marvelon but sometimes a pill containing 50 μg of oestrogen, e.g. Minovlar, is indicated.

A pill containing cyproterone acetate 2 mg combined with 50 μg of ethinyloestradiol (Diane) causes marked improvement in these conditions and is at the same time a contraceptive. In the UK it is marketed only for the treatment of women with these skin conditions, and should not be prescribed primarily as a contraceptive.

Major side effects

Whenever a patient presents with symptoms or signs suggestive of a major side effect relating to pill use, the pill should be stopped immediately and investigation/treatment arranged. Each woman should be warned in advance about these symptoms:

1. Painful swelling in the calf.
2. Pain in the chest or stomach.
3. Breathlessness or cough with blood-stained 'phlegm'.
4. A bad fainting attack or collapse, or focal epilepsy.
5. Unusual headache or disturbance of speech or eyesight.
6. Numbness or weakness of a limb (p. 67, 74).

Circulatory disease: hypertension as a marker

1. All patients on the pill should have their blood pressure (BP) recorded regularly and repeated if necessary to obtain a true record.

2. Repeated recording of diastolic pressures between 80 and 95 mmHg are important *as a marker of the risk of circulatory disease.* This is particularly so if superimposed on other risk factors. For example, a 33-year-old smoker with a BP 140/85 mmHg should stop the pill although a BP of this level would not be significant by itself.

3. Repeated recordings of a diastolic BP above 95 mmHg are usually considered an indication for stopping the pill.

4. Patients with a gradual rise in BP should be changed to a lower dose pill or a progestogen-only preparation. BP levels usually return to normal within three months in either case.

5. Pill-induced hypertension should not be treated with anti-hypertensive drugs.

6. Young women with *essential* hypertension which is well controlled by therapy may sometimes be given a low-dose combined pill under the supervision of a consultant physician (Ch. 5).

Malignant disease

This has been considered under contraindications (pp. 58–60), advantages (p. 61) and disadvantages (pp. 69—73).

Pill users should be told that the statement, 'The pill does not significantly increase the overall risk of developing cancer' has not so far been disproved.

Once a malignant tumour has developed, steroidal contraception should be prescribed only with the approval of the consultant. This rule also applies to benign tumours of the liver.

Indications for stopping the pill

1. Onset of any of the major signs or symptoms listed on pages 58 and 90.

2. A sustained diastolic pressure of more than 95 mmHg or even 85–90 mmHg if other risk factors associated with CVS disease risk are present.

3. Appearance of a new risk factor(s) constituting a risk of CVS disease that is unacceptable for that particular woman. Commonest of these is increasing age in women who smoke (pp. 59, 66, 93.).

4. Onset of jaundice (pp. 58, 88).

5. Before elective surgery — the pill should be stopped six weeks before major surgery and started three to four weeks thereafter provided the woman is ambulant. It should also be stopped for six weeks before and four weeks after completion of treatment for varicose veins whether by surgery or sclerotherapy.

Major *emergency* surgery may be carried out under subcutaneous heparin cover.

The surgeon should always be told if a woman is taking the pill.

It is not necessary to stop POPs before major surgery.

It is not necessary to stop combined pills before minor surgery including sterilisation by laparoscopy.

6. Long-term immobilisation, for example, associated with orthopaedic injury or operation.

7. When pregnancy is desired. The woman should be advised to use additional contraception for at least one month after stopping the pill to allow her to have at least one normal period by which to date the pregnancy. Waiting for three months is recommended by some authorities to allow folate levels to return to normal and theoretically reduce the risk of fetal abnormality. There is no firm evidence to support this advice (World Health Organisation 1981).

8. When contraception is no longer needed.

The first period after stopping the pill is often delayed.

Amenorrhoea for six months after stopping the pill requires investigation.

Breaks in pill taking

To preserve future fertility? Fertility is not enhanced by breaks in pill taking (p. 93).

To reduce side-effects? Most risks and benefits such as venous thromboembolism, and relief from menstrual symptoms apply only while the pill is being taken. There appears to be a duration-of-use effect relating to certain CVS diseases (pp. 65, 66), and for a heart attack the possibility of an ex-use effect related to *previous* duration of use.

Nevertheless, there is no evidence that by taking breaks the risk of developing CVS disease is reduced unless these are long enough to have a real impact on total accumulated years on the pill. In other words, if a pill-user takes a 6 month break every 2 years, she will accumulate 10 years of use in 12 years. This will not, so far as is known, reduce her CVS risk compared with a matched woman taking the pill continuously for 10 years, but it will certainly increase her risk of unplanned pregnancy.

To reduce the risk of cancer? If it is eventually established that OCs do initiate or promote any type of cancer, the harmful effect is likely to be greater with increasing duration of use. Conversely, if the pill reduces a risk, as it does for carcinoma of the ovary and endometrium, the protective effect is greater the longer the pill is used.

This 'swings and roundabouts' effect, with good results of long-term use tending to balance the bad, means that, pending more data, the rules implied in Table 4.6 for minimising circulatory disease risk are also good working rules for cancer risk, provided they are combined with careful screening for cervical neoplasia and breast tumours.

Once again breaks are unlikely to help, as the evidence points to total duration of use.

Duration of use in relation to age. In the present state of our knowledge it seems prudent to set some limits to each of these, considered independently. A suggested scheme is shown in Table 4.6.

Yet, nothing said here overrides the importance of the patient's own views and intuition. If her peace of mind is helped by taking short-term breaks from OCs, she should be helped in the choice of an appropriate short-term alternative.

Table 4.6 Scheme for management in relation to age, smoking and duration of OC use

Management — age and OC use				
Age (years)	30	35	40	45
Smoker	Review	*Change method*	—	—
Nonsmoker		Review	2nd review *(usually change method)*	*Change method*

NB: *Above age 45 (35 in smokers):* Combined OCs contraindicated for contraception (may be usable as therapy). Progestogen-only pill may be a good choice.

Management — accumulated duration of use (pp. 65, 66, 92)				
Duration (years)	5	10	15	20
Smoker	Review	2nd review *(usually change method)*	*Change method*	—
Nonsmoker		Review	2nd Review *(usually change method)*	*Change method*

SPECIAL CONSIDERATIONS

Fertility aspects and amenorrhoea

Pill taking does not cause permanent impairment of subsequent fertility. There may be an initial delay in conception of about three months on average. A few women have amenorrhoea for six months or more which is not related to the type of pill or duration of use and is indistinguishable as regards diagnosis and treatment from other types of secondary amenorrhoea.

There are therefore no grounds for advising regular breaks in OC therapy in order to preserve future fertility.

Secondary amenorrhoea after stopping the pill is not uncommon in women who have had a late menarche, previous episodes of amenorrhoea, very irregular cycles or who are of low body weight, especially associated with anorexia nervosa. Indeed, secondary amenorrhoea may be masked by regular withdrawal bleeds while on the pill.

Women who fail to menstruate after stopping the pill should be investigated in the same way as any other case of secondary amenorrhoea. While the woman is amenorrhoeic the risk of conception is very low and a simple method such as the condom or cap suffices. After menstruation is re-established a very low-dose pill may be prescribed and may even be beneficial in cystic ovary syndromes.

Women with hyperprolactinaemia should use nonhormonal contraception.

Puberty

The average age of the first menstrual period is 11–12 years. Initial cycles are usually anovulatory. An increasing number of girls under the age of 16 are now sexually active and, if they do not have adequate contraception, they run the risk of pregnancy. Both delivery and therapeutic abortion are disastrous at such a young age.

These young girls should be carefully counselled about the implications of embarking on intercourse, particularly in relation to their emotional reactions, the risk of sexually transmitted diseases and of pregnancy.

Regular menstruation should be established before the pill is prescribed. If this rule is followed the risk of subsequent amenorrhoea and cycle irregularity is the same as for girls starting the pill at an older age.

There is no evidence that prescribing the pill for very young girls stunts their growth (pp. 72, 73).

The question of increasing the risk of cancer of any kind in these young girls is still not resolved (pp. 71, 72). Wherever possible, for them the pill of choice is one of the low-dose varieties.

The menopause

The menopause (cessation of menses) usually occurs between 45 and 55 years. The climacteric precedes it by 5–10 years and during this period, although effective contraception is still required, fertility declines. Oestrogen secretion decreases. Although the combined pill compensates for diminishing intrinsic oestrogen production and combats hot flushes and osteoporosis, because of the increased risk of cardiovascular disease, it is not currently recommended for any woman after the age of 45 or after 35 if she smokes.

The progestogen-only pill may be suitable (Ch. 5).

Hormone replacement therapy with oestrogens alone or in part combined with progestogens is not reliably contraceptive.

Pregnancy

Whether taking the pill during early pregnancy increases the risk of congenital abnormality is still not established since other factors such as smoking, alcohol, drugs, X-rays, diet or low social class may contribute to the risk of fetal damage.

Although many 'breakthrough' pregnancies are known to have occurred in women on the pill the incidence of congenital abnormality in this country has not increased since the pill was introduced 20 years ago. Neither of the prospective pill studies in Britain has revealed any increased risk compared with nontakers and a WHO study in 1981 concluded that the risk of teratogenesis is very small.

However, pill taking after conception should always be avoided if possible.

Lactation

Ovulation may be re-established during lactation, particularly if breast feeding is being supplemented or if milk supply is diminishing. Contraception is therefore needed.

The combined pill may impair milk production and is unnecessary. The progestogen-only pill is preferred. When menstruation returns, if maximum effectiveness is important transfer to a combined pill (p. 79).

Patients with intercurrent disease

Reliable protection from pregnancy is often particularly important when other diseases are present. The combined pill, however, could have an adverse effect on the condition. While it is impossible to list every disease, those for which there are data leading to an absolute or relative contraindication to pill use are listed on pages 58–61.

On present evidence, the following conditions do not absolutely contraindicate the use of the pill unless the patient is immobilised — allergies, asthma, Gilbert's disease, multiple sclerosis, myasthenia gravis, Raynaud's disease, renal dialysis, rheumatoid arthritis, sarcoidosis and spherocytosis.

Careful supervision and regular reassessment of whether the patient is still suitable for the pill is essential in all cases.

Diabetes mellitus. The pill may affect glucose tolerance and insulin levels but there is no increased incidence of frank diabetes in OC users.

Insulin requirements may be raised.

Diabetics are particularly prone to hypertension and cardiovascular disease as well as to diseases of the nervous and renal systems and of the eye. The presence of any of these complications absolutely contraindicates the use of the combined pill.

Young diabetics who do not smoke may use the combined pill but for the shortest possible time. The progestogen-only pill is usually preferable, particularly if taken regularly, e.g. at the time of the early evening insulin injection (Ch. 5).

RISKS/BENEFITS

For many women the combined pill provides highly acceptable and reliable contraception. The risks attached to it are real and should not be denied, but their incidence is small (Fig. 4.2) and must be balanced against the benefits of the pill as an effective contraceptive and the protection it affords against many diseases. The risks must also be put in perspective with the other risks women run during their reproductive lives — those of pregnancy, of alternative methods of contraception and attached to life in general.

The safety of the pill can be further increased by:

1. Prescribing it primarily for the 'safer' woman.
2. Ensuring extra care and supervision for those with risk factors.
3. Using pills containing the lowest suitable dose of both oestrogen and progestogen.
4. Careful monitoring of:
   a. Any change in risk factors.
   b. New circumstances, e.g. elective surgery.
   c. Blood pressure.
   d. Headache pattern.
   e. Breasts and cervix, by appropriate screening (Ch 15).

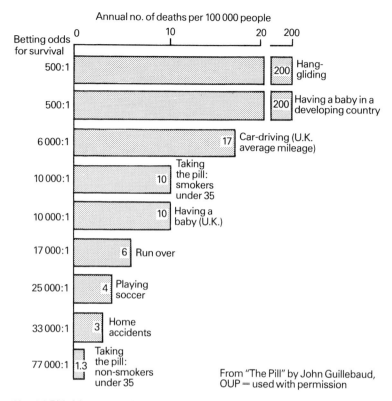

Fig. 4.2 Pill risks compared with other risks women run.

Even more important than following such a scheme is the attitude of the doctor and the nurse and their ability to advise and counsel the patient in a non-directive way. They must not only be conscientious and up-to-date in their knowledge but also relate this successfully to the woman's needs.

Acknowledgement

I would like to express particular thanks to Dr Barbara Law for her valuable assistance in preparing this chapter.

REFERENCES

Back D J & Breckenridge A M 1978 Drug interactions with oral contraceptive steroids. IPPF Medical Bulletin 12 (4): 1–2

Back D J, Breckenridge A M, Orme M 1983 Drug interactions with oral contraceptive steroids. IPPF Medical Bulletin 17 (1): 1–2

Bickerstaff E R 1975 Neurological complications of oral contraceptives. Oxford University Press

Kalache A, McPherson K, Barltrop K, Vessey M P 1983 Oral contraceptives and breast cancer. British Journal of Hospital Medicine 30: 278–283

Lincoln R 1984 F.D.A. Committee· No support for pill–breast cancer link, but cervical cancer connection more ambiguous. International Family Planning Perspectives 10: 27–31

McPherson K, Neil A, Vessey M P, Doll R 1983 Oral contraceptives and breast cancer. Lancet ii: 1414–15

National Association of Family Planning Doctors (Clinical and Scientific Advisory Committee) 1984 Interim guidelines for doctors following the pill scare. British Journal of Family Planning 9: 120–22

Pike M C, Henderson B E, Krailo M D, Duke A, Roy S 1983 Breast cancer in young women and use of oral contraceptives: possible modifying effect of formulation and age at use. Lancet ii: 926–929

Pritchard J A, Pritchard S A 1977 Blood pressure response to estrogen-progestin oral contraceptive after pregnancy-induced hypertension. American Journal of Obstetrics and Gynecology 129: 733–41

Rhodes J M, Cockel R, Allan R N, Hawker P C, Dawson J, Elias E 1984 Colonic Crohn's disease and use of oral contraception. British Medical Journal 288: 595–596

Royal College of General Practitioners 1981 Oral contraception study. Further analyses of mortality in oral contraceptive users. Lancet i: 541–46

Royal College of General Practitioners 1983 Oral contraception study. Incidence of arterial disease among oral contraceptive users. Journal of the Royal College of General Practitioners 33: 75–82

Slone D, Shapiro S, Kaufman D W, Rosenberg L, Miettinen 0 S, Stolley P D 1981 Risk of myocardial infarction in relation to current and discontinued use of oral contraceptives. New England Journal of Medicine 305: 420–4

Stone M, Dent J, Kardana A, Bagshawe K D 1976 Relationship of oral contraception to development of trophoblastic tumour after evacuation of a hydatidiform mole. British Journal of Obstetrics and Gynaecology 83: 913–916

Vessey M P, Lawless M, McPherson K, Yeates D 1983 Neoplasia of the cervix uteri and contraception — a possible adverse effect of the pill. Lancet ii: 930–934

World Health Organisation 1981 The effect of female sex hormones on fetal development and infant health: report of a WHO scientific group. Technical report series 657, WHO, Geneva

FURTHER READING

Briggs M, Briggs M. (eds) Annual research reviews. Oral contraceptives. Eden Press, USA & Churchill Livingstone, Edinburgh

Guillebaud J 1984 The pill, 3rd edn. Oxford University Press

Population Information Programme 1982 Oral contraceptives in the 1980s. Population Reports: June 1982. The Johns Hopkins University, 624 North Broadway, Baltimore, Maryland 21205, USA

Progestogen-only pills

Progestogen-only oral contraceptives, also called continuous progestogen or minipills, contain no oestrogen. They consist of a small dose of one of four progestogens — norethisterone, ethynodiol diacetate, dl-norgestrel or its active isomer levonorgestrel, all of which are 19-nortestosterone derivatives.

The first pills of this type containing chlormadinone acetate were introduced into this country in 1969 but were withdrawn a year later when toxicity tests in experimental animals showed that this progestogen caused breast nodules.

Progestogen-only pills in current use (1984) are shown in Table 5.1.

MODE OF ACTION

The exact mechanisms of action are still unknown (Ch. 3) but are believed to be as follows:

1. Changes in the endometrium making it unreceptive to the implantation of the fertilised ovum. This is probably the prime site of action.

2. Alteration in the cervical mucus rendering it hostile to sperm penetration.

Table 5.1 Progestogen-only pills

Name	Progestogen	Dose	No of pills per packet	Manufacturer
Femulen	Ethynodiol diacetate	500 µg	28	Gold Cross Pharmaceutical
Noriday	Norethisterone	350 µg	28	Syntex
Microner	Norethisterone	350 µg	28	Ortho-Cilag Pharmaceutical
Neogest	Levonorgestrel	37.5 µg*	35	Schering
Microval	Levonorgestrel	30 µg	35	Wyeth
Norgeston	Levonorgestrel	30 µg	35	Schering

*Plus 37.5 µg of inactive isomer.

3. Decreased tubal motility.

4. Ovulation suppression with some progestogens in some cycles.

EFFECTIVENESS

Contraceptive effectiveness varies from one product to another and in studies from different countries. Maximum effectiveness is dependent upon regular pill taking. Reliability depends greatly on the motivation of the woman.

A pregnancy rate of 2–4 per 100 women years is frequently quoted. Higher failure rates occur when patient compliance is poor, but rates as low as 0.5–1 per 100 women years are reported when pregnancies resulting from missed pills are excluded.

Data from the Oxford/FPA study (Vessey et al 1981) show:

1. Overall failure rate = 1.4 per 100 women years. This ranges from 0.5 for women over 35 to 3.0 for those under 25.

2. Little evidence of any relationship between failure rate and duration of use.

3. Marked decline in the risk of accidental pregnancy with age.

Although based on small numbers (796 woman years) these results are very encouraging and show that progestogen-only pills are the second most effective method of reversible contraception. They also refute the suggestion that there is an increase in the pregnancy rate with duration of use. There is no evidence that a resistance to its contraceptive effect develops after a period of time.

INDICATIONS

1. *For women who wish to use oral contraception but*
   a. are not prepared to run the possible thrombogenic and undesirable metabolic effects associated with oestrogens;

b. are unable to tolerate oestrogens or develop oestrogen-related side-effects on the combined pill;

c. for whom oestrogens are contraindicated.

2. *For older women.* It is now well established that women over the age of 35 on the combined pill run a greater risk of dying from cardiovascular disease particularly if they smoke (p. 66). The progestogen-only pill provides an alternative contraceptive for this group.

3. *During lactation.* Amenorrhoea, which is fairly common in progestogen-only pill takers, is of no concern to this group who expect the return of menstruation to be delayed while they are breast feeding.

Progestogen-only pills do not interfere with the quantity of milk produced. Small changes occur in the relative concentration of protein, lactose and lipids in the milk but are of little consequence. Immunoglobulins, which provide antibodies to the infant, are not affected even when these pills are started immediately after delivery when colostrum, rich in immunoglobulins, is being secreted.

Although a small amount of progestogen may be ingested by the baby, there is no evidence of cumulation or suggestion that it has any short-term deleterious effects. However, as the possible long-term effects have not yet been studied, some manufacturers advise prescribing their products 'with caution' for breast-feeding mothers. No such warning is given for the other preparations. Workers in Sweden using 30 μg levonorgestrel (Nilsson et al 1977) have shown that 0.1% or less of the dose given to the mother appears in the milk. The consensus of opinion at present is that progestogen-only pills provide a very safe and effective contraceptive for breast-feeding mothers.

4. *For women with a history of oligomenorrhoea or late menarche.* There is some evidence that after the combined pill is stopped the return of fertility may be delayed (p. 93). Some doctors prefer to prescribe progestogen-only pills for these women. However, as the problem usually presents in young girls, for whom effectiveness is all-important, the advantages of the combined pill will often make it the method of choice. The possible risks to future fertility attached to taking the combined pill must be weighed against those of termination of pregnancy, should that be resorted to if the minipill fails.

5. *For women with diabetes mellitus.* There is evidence that the combined pill carries a greater risk of coronary and cerebrovascular disease in diabetic women, and, although the incidence of retinopathy in pill takers is not increased, it seems to progress unusually rapidly.

These problems do not arise with progestogen-only pills, which may therefore be the method of choice for this group.

6. *For women with mild hypertension or where the blood pressure is well controlled.* In these cases the approval of the consultant should first be sought.

CONTRAINDICATIONS

Although a long list of conditions contraindicating the use of progestogen-only pills is found in the manufacturers' data sheets, these warnings are often precautionary with no epidemiological evidence to support them. There are, however, some situations where it is unwise to prescribe progestogen-only pills.

Absolute contraindications

1. Known or suspected pregnancy.
2. When complete protection against pregnancy is essential.
3. Menstrual irregularity of uncertain cause. Taking the minipill will confuse the picture and make diagnosis more difficult.
4. Previous ectopic pregnancy and those who are at higher risk, e.g. with a history of pelvic inflammatory disease (PID) or tubal surgery. Progestogen-only pills may increase the risk of ectopic pregnancy (p. 104).
5. After a hydatidiform mole, until the urine is free of human chorionic gonadotrophin.
6. Malignant disease of the breast. Although there is no evidence that progestogen-only pills affect the prognosis, doctors are still cautious about prescribing them.

Relative contraindications

1. Those who find irregular menstrual bleeding unacceptable either for social or for cultural reasons.
2. A history of thromboembolic disease. Although there is no evidence that progestogen-only pills increase the risk of recurrence, manufacturers still advise against prescribing them for patients with this history. Yet some of these patients will often refuse to use any contraceptive other than the pill. Provided the risks and benefits of progestogen-only preparations are thoroughly explained to them, these pills may be the best method of preventing unwanted pregnancy in such cases.

3. Severe liver disorders with persistent biochemical change.

4. Malabsorption syndrome.

5. Drugs which may interact (p. 111).

6. Opinions vary on the importance of previous cone biopsy of the cervix. Some suggest that since a proportion of the cervical glands have been excised, the contraceptive effect on the cervical mucus will not operate, but there is no scientific evidence to support this theory.

ADVANTAGES

Exhaustive laboratory investigations have revealed no serious adverse effects. Subjective side-effects are notoriously difficult to assess and different observers report widely differing incidence. However, they seldom present major problems (Fotherby 1982).

There are few data available from large epidemiological studies such as for the combined pill but a further report from the Oxford/FPA contraceptive study confirms the overall reassuring picture which is emerging.

The main advantages are:

1. Contraceptive effectiveness is second only to the combined pill.

2. The pill-taking regimen is easy to follow.

3. Excessive weight gain, nausea and headaches are not a problem.

4. There is little effect on libido.

5. Premenstrual tension, dysmenorrhoea and mastalgia are often relieved.

6. There is minimal alteration in carbohydrate and lipid metabolism. High density lipoprotein (HDL) levels are not significantly altered (Briggs & Briggs 1983).

7. Liver function tests are not affected.

8. Adverse effects on clotting mechanisms are only minimal. It is important, however, to remember that in itself this is not proof that minipills will be devoid of all thrombotic risks. It is reasonable nonetheless to believe that the thrombotic risk is small. Disturbances of clotting mechanisms produced by the combined pill are quickly reversed when progestogen-only pills are substituted.

9. There is no consistent effect on blood pressure. When prescribed for patients who have become hypertensive on the combined pill, the blood pressure often drops to normal.

10. There is no epidemiological evidence of increased risk of cardiovascular or malignant disease.

11. Subsequent fertility is not impaired.

12. No harmful effects result from an overdose, even when taken by a child.

DISADVANTAGES

1. Alteration of the menstrual pattern is the main problem. Unfortunately one cannot predict those who will experience menstrual pattern change or how severe the change will be.
   a. Spotting and breakthrough bleeding are common.
   b. The duration and volume of flow may change.
   c. The length of cycles varies widely, but regularity improves with duration of use. Irregular short cycles are particularly common in the first few months with spells of amenorrhoea developing later on. On average, 70% of cycles will last 28 ± 5 days, 20% less than 23 days and 10% more than 30 days. Cycle irregularity is the commonest reason for women abandoning the pill. Yet some feel so 'positively well' on it that they are prepared to accept this inconvenience. Irregular cycles are often anovulatory and the risk of pregnancy consequently nil.

2. In regular cycles ovulation is probably not inhibited and it is often in such women that the risk of pregnancy is greatest.

3. A small number of women may develop luteal cysts.

4. Pregnancies occurring in minipill users are more likely to be ectopic than in women not using contraception. Ectopic pregnancy occurs in 1:1000 users. In one series 4.3% of the 73 unplanned pregnancies were ectopic — a rate similar to that reported with IUDs.

CLINICAL MANAGEMENT

Assessment

Suitability for the progestogen-only pill is based on history and physical examination.

History

A history should be taken particularly regarding age, the need to avoid pregnancy, motivation and those conditions which might contra-indicate progestogen only pills use.

Examination

1. Record blood pressure.
2. Note weight.
3. Perform pelvic examination.

Although it is reassuring to the doctor to establish that the pelvic organs are normal, many women dislike 'internal examinations'. Pelvic examination should therefore not be insisted on provided the patient gives a normal menstrual history and the date of a normal last menstrual period is clearly established to exclude pregnancy. Pelvic examination can then be delayed until a later date.

4. Take a cervical smear if indicated by clinic or practice screening policy for women in that age group (Ch. 15).
5. Examine breasts to exclude disease, particularly in older women, and in younger women, according to clinic or practice screening policy (Ch. 15).

Choice of pill

1. One cannot be dogmatic about which pill to choose for any individual patient as any of the six products now available may prove equally acceptable. Often it is a case of trial and error.

2. Results vary with different progestogens, different doses of the same progestogen, ethnic group, body weight and of course motivation.

3. There is no proof that one pill is more effective than another.

4. Incidence of menstrual irregularity is not much influenced by the particular progestogen, although norethisterone seems to be associated with a smaller proportion of short cycles and less variation in cycle length than levonorgestrel.

5. Less steroid appears in the breast milk with pills containing a small dose of levonorgestrel and these preparations may therefore be considered the best choice during lactation.

6. Some women seem to develop symptoms such as headaches, which may be due to a sensitivity to the progestogen, but unfortunately there is no way of predicting this.

7. Very often the best pill for any individual is the one she asks for because she has heard good reports of it from her friends!

Administration

1. The regulations covering the prescription of progestogen-only pills correspond to those for the combined pill (pp 76 and 77).

2. The pill-taking regimen should always be carefully explained. No matter how explicit the patient information leaflet contained in each packet may be, it does not replace careful instruction and explanation from the doctor or the nurse. This is particularly important when changing from the cyclical regimen of combined pills to the continuous regimen of progestogen-only preparations.

3. All the pills are dispensed in bubble packs with the day of the week clearly marked as an 'aide memoire'. The number of pills in each packet varies with different brands (Table 5.1).

4. The first pill is taken
  a. when starting for the first time — on the first day of the next period;
  b. when changing from a combined pill — on the day following the last pill in the packet;
  c. after delivery — either between 7 and 28 days later or as soon as lactation is fully established. It is not necessary to await the return of menstruation;
  d. after termination of pregnancy — on the same day as the termination or the day after.

5. One pill is taken every day thereafter without a break whether menstruation occurs or not.

6. Tablets should be taken at the same time each day and for most women the best time is between 6.00 and 7.00 p.m. Since it takes about 4 hours for the maximum effect on the cervical mucus to be achieved, taking it at this time means that it is at its most effective by the time the majority of people have intercourse. Each woman will have to work out the optimal time for herself. The important thing is that the pills should be taken regularly several hours before intercourse usually takes place.

7. Extra contraceptive precautions for *14 days* are recommended:
  a. for the first 14 days of pill-taking irrespective of the circumstances in which it is started;
  b. if the pill is taken more than 3 hours late;
  c. if a pill is forgotten — a tablet should be taken when the mistake is recognised and then the regular pill for that day taken at the usual time;
  d. if vomiting occurs within 3 hours of pill-taking — another tablet should be taken. If this too is vomited, consider that a pill has been missed and advise additional contraception for 14 days *after* the symptoms have subsided;
  e. if an attack of severe diarrhoea occurs. The pill may not be properly absorbed. Extra precautions are required for 14 days

after the diarrhoea has stopped, the pill being taken as usual in the meantime;

f. when changing from a combined pill.

Many doctors now question the need to advise extra precautions for 14 days in these circumstances and believe that to do so for 48 hours would be sufficient. However, this practice has not, so far, been properly evaluated.

8. When transferring from a progestogen-only pill to a combined pill begin on the first day of the next menstrual period. During lactation, menstruation may not yet have been re-established. In that case the first combined pill should be taken on the day after the last progestogen-only pill.

Follow-up

1. The patient should be seen three months after starting the pill and at six-monthly intervals thereafter.

2. Where no problems arise routine examinations should be kept to a minimum.

a. Routine weighing is unnecessary but if the patient believes she is losing or gaining weight this can easily be checked against the figure recorded at the first visit.

b. Blood pressure should be recorded after three months and yearly thereafter.

c. Routine breast examinations are unnecessary for healthy young women just because they are taking progestogen-only pills, but yearly clinical examination of the breast is good clinical practice for all women over the age of 35. The consultation also provides an excellent opportunity to teach the woman self-examination of the breasts (Ch. 15).

d. Pelvic examination after any arbitrary period of time is not necessary if the patient is not complaining of any problems, if cycle control is good and if there is no suspicion of pregnancy.

e. Cervical smears should be repeated according to the screening programme guidelines (Ch. 15).

COMPLICATIONS AND THEIR MANAGEMENT

1. Women will sometimes complain of vague symptoms which they fear might be caused by the pill. With support and reassurance these symptoms soon disappear.

2. Minor side-effects such as headaches, nausea and breast tenderness often subside after the first few months, just as they do on the combined pill. If they do not, a different progestogen should be tried. There are no rational guidelines as to which progestogen to choose.

3. Menstrual irregularities pose the biggest problem, particularly among women of low body weight.

If irregular periods occur, check that the patient is taking the pills regularly and exclude malabsorption due, for example, to vomiting or diarrhoea. Enquire about other drugs (p. 111). Information on whether or not drug interaction occurs with progestogen-only pills is inadequate — at present the possibility that it may be responsible for breakthrough bleeding must be accepted.

If poor cycle control persists, another pill can be tried. Changing to a more potent progestogen is often suggested but in practice prescribing a less potent preparation often works just as well. The potency of different progestogens is discussed in Chapter 3.

Oestrogen should *not* be prescribed to control cycle irregularity since it interferes with the contraceptive effect of progestogen on the cervical mucus. In older women irregular cycles around the time of the menopause should never be accepted as being due to the pill. The pill should be stopped, menstrual pattern assessed and, if irregular bleeding persists, gynaecological investigation carried out.

Amenorrhoea occurs for more than 60 days in 2–5% of patients during the first year of pill-taking. Pregnancy should be excluded and this is particularly important if pills have not been taken regularly or possibly not absorbed. The pills should be stopped immediately if pregnancy is confirmed.

During the course of long-term medication, amenorrhoea can last for many months. It is not in any way harmful. Some women accept it, happy to be rid of the monthly period. Others, constantly fearing pregnancy, pay repeated visits to the doctor for pregnancy tests to put their minds at rest. They should be reassured that they are well protected against pregnancy. Changing to another preparation may solve the problem. The only solution for those who are worried by persistent amenorrhoea is to abandon the pill.

Periods generally return to normal quickly after the minipill is stopped.

Amenorrhoea continuing for six months after the pill has been stopped should be investigated and treated in the same way as amenorrhoea in a woman who has never taken hormonal contraception.

For amenorrhoea in older women see below.

4. Lower abdominal pain is occasionally a problem. Causes not related to contraception should be excluded. Two possibilities must always be borne in mind; ectopic pregnancy, particularly if the period is a few days late; torsion or rupture of a functional ovarian cyst. In these cases referral for ultrasound and/or laparoscopy is often needed to establish a diagnosis.

SPECIAL CONSIDERATIONS

Adolescents

Minipills are not generally recommended for teenagers for whom effective contraception is usually of prime importance. For them even the small failure rate is not acceptable (p. 107). However, if the combined pill is contraindicated the minipill may be a better alternative than the IUD, which is now recognised as carrying a significant risk of impairment of future fertility.

Older women

It is now generally accepted that the combined pill is unsuitable for women over the age of 44, except in exceptional circumstances, and for women over the age of 35 who smoke. The minipill is becoming increasingly popular for them. However, it cannot be recommended unreservedly for older women who are also heavy smokers, until further evidence of its safety is obtained. Recent reports that Micronor, Noriday, Microval and Norgeston reduce high-density lipoprotein levels only minimally are reassuring. So too is the evidence that progestogen-only pills have very little effect on the blood pressure.

A problem is how to know when the menopause is reached. The age at which the patient's mother or sister stopped menstruating is helpful and symptoms such as hot flushes can also act as a guide. No specific test is available. However, there is some evidence that, in progestogen-only pill users with complete amenorrhoea, a high level of follicle stimulating hormone (FSH) is diagnostic of the menopause. If this is confirmed it will prove a most useful test. Until then doctors are advised to tell women in their 40s who develop prolonged amenorrhoea on the minipill to stop it and use alternative contraception for at least another year. If menstruation is re-established the pill can be restarted.

Future fertility

This is not impaired. The woman's fertility should be the same as if she had never taken the pill, provided she has escaped the very small risk of ectopic pregnancy.

Women with amenorrhoea on the pill are at no greater risk of infertility than those with regular cycles or irregular bleeding. In this context it is important to remember that 10–15% of all couples in the UK are involuntarily infertile.

Medical disorders

Patient information leaflets list a number of medical conditions in which progestogen-only pills are contraindicated but offer little evidence to support the opinion. It is often a counsel of caution. Each doctor will have to decide for himself whether or not to prescribe this pill for these patients.

1. A history of superficial thrombophlebitis is listed as a contraindication yet many studies have confirmed that the minipill confers no increased risk of recurrence of this condition and no increased risk of deep venous thrombosis. One has always to weigh up the risk of recurrence in pregnancy or in the puerperium, should a less effective method be chosen and fail.

2. The question of prescribing the progestogen-only pill for women with breast cancer is still unresolved. Some authorities argue that, as progestogens are used to treat the condition in advanced stages, they are not contraindicated for women with the disease. However, in the present state of our knowledge it would be wise to avoid hormonal contraception for hormone-dependent cancers.

3. It is not contraindicated in women with benign breast disease. On the other hand there is no evidence that it confers the same protection against it which has been demonstrated with combined preparations (p. 61).

4. It may be prescribed for women with atypical cervical smears, Raynaud's disease, anorexia nervosa and for those who wear contact lenses.

5. Chloasma induced by the combined pill can fade when the woman is changed to the minipill.

Interpretation of laboratory tests

Progestogen-only pills do not significantly alter any of the common biochemical tests.

Pregnancy

1. *Prior to conception.*

There is no evidence that women who become pregnant immediately after stopping the progestogen-only pill are at increased risk of fetal abnormality. There is no justification for advising them to use alternative contraception for two months or any other arbitrary period before trying to conceive.

2. *During pregnancy.*

No matter how careful one is to avoid prescribing any contraceptive pill during pregnancy, occasionally a patient may continue to take it after she has conceived. There have been no reports of any significant increase in birth defects in women becoming pregnant while taking the minipill. These women should be reassured that they will have normal pregnancies and normal babies.

Lactation (p. 101)

Drug interaction

Interaction between progestogen-only pills and other drugs with resultant reduction in contraceptive cover appears to be minimal. The possibility, however, should be borne in mind when considering progestogen-only pills for women on long-term therapy with rifampicin, antibiotics, some sedatives, anti-rheumatic preparations and most anti-epileptic drugs except Epilim (sodium valproate), with which interaction does not seem to occur. Additional precautions should be advised during and for 14 days following short courses of antibiotics.

If doctors are suspicious that drug interaction is occurring between progestogen-only pills and any other drug they prescribe, it is important that they should notify the Committee on Safety of Medicines on the appropriate yellow card.

Before major surgery

There is no evidence that it is necessary to stop progestogen-only pills for four to six weeks before major surgery as is advised with combined preparations (p. 91).

Indications for stopping the pill

1. Pregnancy.
2. Metrorrhagia which persists in spite of changing preparations.
3. Blood pressure which has risen above 160/100 mmHg.
4. Acute attack of liver disease.
5. A first attack of *very severe* headache with or without visual disturbances, or any other acute episode that could be interpreted as resulting from a transient ischaemic attack.
6. Signs or symptoms, e.g. chest pain, suggestive of a cardiovascular incident.
7. When another pregnancy is desired. The pill should be stopped on the first day of a period.

When the pill is stopped and the patient does not want to become pregnant, another effective contraceptive should be substituted.

RISKS/BENEFITS

In selecting any contraceptive for a patient the risks and benefits have to be carefully worked out. Compared to combined pills, progestogen-only pills are associated with

1. higher pregnancy rate,
2. higher discontinuation rate,
3. more menstrual irregularity,
4. greater risk of ectopic pregnancy.

On the other hand

1. they have no major undesirable side-effects,
2. mortality associated with their use has never been reported in this country.

For some women the small risk of failure makes them unacceptable. Others, particularly those who just plan to delay a pregnancy, are happy to take this chance in exchange for the benefits conferred. The combined pill, although the most effective of all reversible contraceptives, does carry the hazards associated with oestrogen. A less effective method free from those risks appeals to many.

Older women, whose waning fertility reduces the risk of failure and in whom the risks attached to the combined pill are greatest, are particularly suitable for progestogen-only therapy.

Failure rates with progestogen-only pills and correctly inserted IUDs are about the same but, although menstrual irregularities may be a problem with both methods, menorrhagia seldom occurs in these pill users. Nor is there increased risk of pelvic inflammatory disease.

Progestogen-only pills should also be considered as an alternative to female sterilisation, which although more effective, does carry definite risks both during the operation and thereafter (Chamberlain & Brown 1978) (Ch. 10). The risk of ectopic pregnancy after sterilisation is also greater than that associated with the progestogen-only pill (Tatum & Schmidt 1977).

Progestogen-only pills are more effective than mechanical contraception and their convenience in comparison to the cap or condom appeals to many women.

Ex-users are at no known increased risk of any kind.

In the search for a better contraceptive the minipill has not been fully exploited. It has not even been very popular. In 1978 only 2% of oral contraceptives purchased from chemists in Great Britain were progestogen-only pills. By 1984 the proportion had risen to only 8%. In the past 10 years enough information about and experience with progestogen-only preparations has been gained to encourage doctors to prescribe them for many more women as a first choice rather than as a last resort.

REFERENCES

Briggs M, Briggs M 1983 Plasma lipids in women using progestogen-only oral contraceptives. British Journal of Obstetrics and Gynaecology 90: 549–552
Chamberlain G, Carron-Brown J (eds) 1978 Gynaecological laparoscopy: working party report. Royal College of Obstetricians and Gynaecologists, London
Fotherby K 1982 The progestogen-only contraceptive pill. British Journal of Family Planning 8: 7–10
Nilsson S, Nygren K G, Johansson E D B 1977 d-norgestrel concentrations in maternal plasma, milk, and child plasma during administration of oral contraceptives to nursing women. American Journal of Obstetrics and Gynecology 129: 178–184
Tatum H J, Schmidt F H 1977 Contraceptive and sterilisation practices and extrauterine pregnancy: a realistic perspective. Fertility and Sterility 28: 407–421
Vessey M P, Yeates D, Flavel R 1981 Progestogen-only oral contraceptives. In: Proceedings of Symposium on Current Fertility Control 1–12, Royal College of Physicians 1979. John Wyeth and Brother Ltd, London

Injectable contraceptives

Types
Depot medroxyprogesterone acetate
Norethisterone oenanthate
Other injectable steroids

Mode of action

Dose and effectiveness

**Availability and status in the world
and UK**

Indications

Contraindications

Advantages

Disadvantages

Clinical management
Assessment
Choice of preparation
Administration
Patient information
Follow-up
Complications and their management

Risk of carcinogenesis

Risks/benefits

The discovery that esterification of a progestogen alcohol produced a long-acting drug when injected was made in 1953. In 1963 the Upjohn Company began clinical trials of medroxyprogesterone acetate injections (DMPA: Depo-Provera). Schering AG had begun trials of norethisterone oenanthate (NET OEN) in 1957.

Experience with these drugs dates back to the early 1960s. Depo-Provera has been used as a contraceptive by over 10 million women around the world and has been studied clinically for over 20 years. It has been used far more extensively and for longer than NET OEN. Consequently much more has been published and is known about it. The controversies surrounding it are therefore all the more surprising as a close investigation of the benefits and risks does not appear to warrant the protracted and bitter opposition to its use, which emanates largely from lay sources.

TYPES

Depot medroxyprogesterone acetate (Depo-Provera, *Upjohn*)

A derivative of 17αhydroxyprogesterone in aqueous microcrystalline suspension: 50 mg/ml — 1,3,5 ml vials.

Norethisterone oenanthate (Noristerat, Norigest, *Schering*)

A derivative of 19-nortestosterone: 200 mg/ml in a vehicle of benzyl benzoate and castor oil — 1 ml vials.

Other injectable steroids

1. Monthly combined oestrogen/progestogens: good cycle control but a tendency to steroid accumulation.
2. Biodegradable microcapsules: progestogens or oestrogen/progestogen combinations released by slow diffusion.
3. Subcutaneous implants

These are all technically possible and some have been tested in humans. However, there are still many difficulties to be overcome before they can be considered realistic alternatives to present contraceptive methods.

MODE OF ACTION

Inhibition of ovulation at hypothalamic level with abolition of cyclic LH, FSH and oestradiol secretions.

Contraceptive effects on luteal and tubal function, endometrium and cervical mucus (Ch. 3).

DOSE AND EFFECTIVENESS

DMPA

Manufacturers recommend 150 mg (3 ml) given within the first 5 days of the menstrual cycle and repeated every 90 days (or 12 weeks). Many other regimens have been tried and proved effective, e.g. 250 mg every 16 weeks, 450 mg every 24 weeks.

Pregnancy rate. 150 mg every 12 weeks — 0.0–1.2 per 100 women years. 450 mg every 24 weeks — 0.49 per 100 women years. The latter regimen has a higher continuation rate than the former.

NET OEN

200 mg (1 ml) given within the first 5 days of the menstrual cycle, repeated every 8 weeks for the first 6 months and thereafter at 12 weekly intervals was formerly recommended but an *8 weekly regimen* continued long term is now advised.

Pregnancy rate. 200 mg every 8–12 weeks — 0.01–1.3 per 100 women years.

AVAILABILITY AND STATUS IN THE WORLD AND UK

DMPA

Use has been re-endorsed by the World Health Organisation (WHO) (1978) and the International Planned Parenthood Federation (IPPF) (1980). It is currently available for long-term use in more than 80 countries, which include Sweden, Germany, Denmark, Netherlands and New Zealand. The Food and Drug Administration in the USA has vetoed its use in spite of several recommendations by its own Obstetrical and Gynaecological Subcommittee.

DMPA has been licensed for short-term use as a contraceptive in the UK for many years. In April 1982 the Committee on the Safety of Medicines (CSM), after considering the matter in depth for several years, recommended that it should be licensed for long term use 'in women for whom other contraceptives are contraindicated or have caused unacceptable side effects or are otherwise unsatisfactory'. However, for the first time since the inception of the CSM the Minister of Health chose to over-rule the Committee's advice.

The Upjohn Company appealed to an independent panel, a procedure laid down by the Medicines Act (1968). The panel recommended a long-term licence in substantially the same terms as those advised by the CSM and to this the Minister of Health agreed in April, 1984. This long awaited decision must be welcomed by all doctors who wish to offer their patients the broadest possible spectrum of contraceptive advice.

NET OEN

Available for use in Germany and in third world countries such as Mexico, Peru, Zimbabwe and Togo. It is now marketed in the UK, as a product licence for antifertility use has been granted.

INDICATIONS

Until recently both DMPA and NET OEN were only licensed for short-term use in the following circumstances:

1. In conjunction with rubella immunisation.

2. For the partners of men undergoing vasectomy (two injections and possibly more at three-monthly intervals will be required). In addition, it was customary to include postpartum women awaiting interval sterilisation.

It is probable that DMPA (and possibly NET OEN, although less is known about it) carries a lower risk of death and a lower morbidity rate than many other forms of contraception such as the combined pill and the IUD. It should therefore be available to all women on the same basis as these other methods.

It is particularly suitable for:

1. Unreliable pill takers (for whatever reason).

2. Women in whom oestrogens are contraindicated and maximum protection is required, e.g. older women, lactating women.

3. Women in whom long-term progestogens are indicated, e.g. premenstrual and menstrual problems (including postmenarcheal girls and menopausal women).

4. Patients suffering from homozygous sickle cell disease. A case-controlled crossover trial demonstrated highly significant improvement in the haematological picture and a reduced incidence of painful crises in patients treated with 3-monthly injections of 150 mg DMPA. The authors suggest that this may be the method of choice in such sufferers (Ceulaer et al 1982).

CONTRAINDICATIONS

Pre-existing malignant disease in breasts or genital organs is generally regarded as a contraindication to hormonal contraception. However, DMPA has been used in the treatment of many of these malignancies and is very widely used in the treatment of endometrial cancer in amounts greatly in excess of the contraceptive dose.

Absolute

1. Pre-existing pregnancy. However, there are no convincing reports of permanent fetal abnormalities even when the mother has conceived during treatment or been given DMPA when already pregnant.

2. Women wishing to conceive immediately after the effective duration of the proposed injection.

Relative

1. Women who are unlikely to accept menstrual irregularities, especially amenorrhoea, or who may have religious or cultural taboos associated with bleeding.

2. Puerperal administration, especially in the first week after delivery, increases the likelihood of heavy and prolonged bleeding. Delaying the first injection until five or six weeks postpartum improves the bleeding pattern.

3. Lactation is not inhibited and may actually be enhanced, but, as minute traces of both DMPA and NET OEN are excreted in breast milk, it is best to delay administration until at least six weeks postpartum. By this time there is no possibility of an adverse effect on the neonatal hypothalamus or liver.

4. During investigation of carbohydrate metabolism.

ADVANTAGES

1. Its use/effectiveness overall is probably higher than that of the combined pill.

2. Continuous motivation is not essential because of the long duration of action of a single injection.

3. Administration is easy and independent of coitus.

4. There are no oestrogenic side effects.

5. Lactation may be enhanced due to increased production of prolactin (Fraser & Weisberg 1981).

6. Amenorrhoea may be medically advantageous, e.g. in a woman with iron-deficiency anaemia.

7. Premenstrual and menstrual symptoms may be relieved. This may be particularly helpful in cases of premenstrual aggression, mood swings and increased epileptic attacks, especially in the mentally handicapped.

8. Reduction in the incidence of moniliasis and pelvic inflammatory disease associated with gonococcal infection reported with DMPA.

9. Certain pathological conditions may be ameliorated, e.g. hirsutism, endometriosis and homozygous sickle cell disease.

10. There is no effect of any significance on blood coagulation and fibrinolysis, on blood pressure or on the liver. Patients with primary biliary cirrhosis or chronic active hepatitis may even improve.

11. Both NET OEN and DMPA may stimulate erythropoiesis. Significantly improved haemoglobin levels have been shown with NET OEN.

12. Acceptability is high, especially among those whose friends are happy using it.

13. Continuation rate is better than with oral contraceptives (Department of Medical and Public Affairs 1975).

DISADVANTAGES

1. Impossibility of ceasing treatment once injection has been given.

2. Administration by injection is unacceptable to some.

3. Menstrual disturbances are the rule and normal cycles the exception. They are one of the main reasons for discontinuation.

a. DMPA: bleeding may be frequent and irregular, although it is usually not heavy. It tends to become less frequent with duration of use, so that by the end of 1 year 35% of women will have had complete amenorrhoea during at least 1 injection cycle. Heavy prolonged bleeding is more likely if the drug is given in the puerperium.

b. NET OEN: the general pattern is similar but one study claims a higher proportion of 'normal' cycles and there is some evidence that amenorrhoeic cycles become less frequent with duration of use (Howard et al 1980). Return of regular menstruation is probably quicker than with DMPA, in association with earlier postinjection ovulation.

4. Weight gain.

a. DMPA: most women, but not all, gain weight (0.5–2 kg in the first year) and some continue to do so. A few become noticeably and increasingly obese, although already fat women seem less likely to gain weight than thin ones. This weight increase is almost certainly due to increased appetite and is not associated with fluid retention. Many women complain of abdominal distension and 'feeling bloated'. Increased breast activity may be associated with tenderness, heaviness and even the need for a bigger brassiere.

5. Delayed return of fertility.

a. DMPA: pregnancy is very unlikely for 8 months after the last injection but the pregnancy rate then rises rapidly so that 75% of women who try to, will conceive by 18 months and 95% by the end of 2 years.

b. NET OEN: the quicker rate of return of ovulation suggests that the delay in conception will be less, but good studies are lacking.

There is *no evidence* that either has a *permanent* effect on fertility.
6. Adverse effects on the fetus.

   a. DMPA: these are extremely rare. However, transient enlargement of the clitoris has been reported in three neonates whose mothers had received DMPA. More severe masculinisation has been reported in women treated with medroxy-progesterone for threatened or habitual abortion (Schardein 1980) although derivatives of 17αhydroxyprogesterone are less androgenic than those of 19-nortestosterone.

   b. NET OEN: masculinisation of the female fetus has been reported when 19-nortestosterone derivatives were given in the first 12 weeks of pregnancy in the treatment of threatened or habitual abortion. Hence the recommendation to give the initial dose within the first 5 days of the cycle. No cases have been reported in contraceptive failure with NET OEN.

7. Subjective effects. These are legion with all types of hormonal contraception and even with IUDs. There is no reliable evidence that the incidence of headaches, loss of libido, mood changes, dizziness, etc., is any higher with either preparation than with these other forms of contraception.

8. Metabolic effects

   a. DMPA: Carbohydrate metabolism: there may be an exaggerated insulin response to glucose tolerance test and therefore carbohydrate metabolism in diabetic women may be affected. Lipid metabolism: there is a small reduction in HDL-C with four times the contraceptive dose but DMPA probably has less effect than other progestogens. Fluid/nitrogen balance: no effects demonstrated.

   b. NET OEN: much less information is available. What there is shows a parallel pattern. Total cholesterol, triglycerides, lipoproteins, glucose tolerance, plasma factor X and anti-thrombin III do not appear to be affected. Levels of HDL-C are moderately but significantly decreased although this effect does not become more marked with duration of use.

9. Systemic effects

   a. Galactorrhoea may occur in those who are not breast feeding.

   b. Mild androgenic effect is more than counteracted by its antiandrogenic action.

   c. Significantly raised haemoglobin concentration and other haematological indices have been shown in women receiving NET OEN irrespective of their menstrual bleeding pattern.

10. Other effects: enuresis may recur in women who were enuretic

in adolescence. This may be associated with the relaxing effect of progestogens on smooth muscle.

CLINICAL MANAGEMENT

Assessment

Medical history

It is good clinical practice to take a medical history but there are no special areas of concern.

Examination

1. Record the weight — more important than with other hormonal methods.

2. Record the blood pressure — desirable but, on present evidence, probably not relevant to the method.

3. Examine the breasts — desirable to exclude mammary carcinoma at the first visit: otherwise according to clinic or practice screening policies.

4. Carry out pelvic examination — good preventive medicine but only relevant to the method in the late puerperium to exclude the possibility of retained products, and to exclude pre-existing pregnancy when there are other indications that this may have occurred. May be counter-productive if it lessens patient acceptability and prevents long-term continuation.

5. Cervical smear — according to clinic or practice screening policies.

Choice of preparation (Table 6.1)

Both are now licensed and marketed in the UK so the choice should be made on clinical grounds.

Administration

Both preparations are given by deep intramuscular injection into the gluteal region, or into the deltoid in the very obese.

1. DMPA: if there is any sediment present, the vial should be shaken well before charging the syringe.

Table 6.1 Comparison of depot medroxyprogesterone acetate and norethisterone oenanthate

	DMPA	NET OEN
Injection interval	12–16 weeks (up to 24)	8 weeks
Ease of administration	Easy, almost painfree	Slightly more difficult (oily) Some discomfort
Effectiveness	0–1.2 per 100 women years	0.01 per 100 women years
Menstrual disturbances	Considerable	Similar but more rapid return to normal
Return of fertility	Mean delay of about 1 year	More rapid
Weight gain	Usual, sometimes considerable	Occasional
Fetal risk	Theoretical only	Masculinisation possible if given in first 16 weeks of pregnancy

The injection should be given preferably within the first five days of menstruation.

2. NET OEN: much easier to handle at body than at room temperature. It must be given within the first five days of menstruation.

Patient information

It is essential to discuss the possible side-effects of these preparations with each woman before they are administered. Failure to do this with postpartum recipients and the subsequent anxieties raised by abnormal bleeding patterns have been the main source of adverse publicity in the UK.

Topics which must be discussed include:

1. Amenorrhoea. This is likely. An explanation about the functions of menstruation, reassurance that a 'monthly clean out' of the 'bad blood' is not necessary for good health (and does not require a D & C for its removal) may make all the difference between happy acceptance and chronic anxiety and dissatisfaction (Wilson 1976).

2. Frequent irregular bleeding. Possible, but not usually heavy. Reassurance that medical advice is available if it occurs is important.

3. Weight gain. Likely with DMPA, but not inevitable as it is associated with increased appetite. Skinny women (often heavy smokers) sometimes ask for DMPA as a means of increasing their size (especially of their bust) but it may not be effective in this respect.

4. Delay in fertility return. Women should be told that on average it takes a year from the time of the last injection to conceive. It is quite wrong to give it postpartum to a woman who wants to conceive again within the next nine months.

5. Long-term risks. How much the doctor tells the individual patient about the theoretical long-term risks must depend on his own judgement of the patient's intelligence and background. 'Informed consent' has as many meanings as there are doctors and patients but some attempt should be made to explain the present licensing regulations and the requirement by the Minister of Health that all recipients must be notified to a central registry so that long term monitoring is possible. An information leaflet, agreed between the company and the Minister, is available and should be given to each new potential recipient so that she can read it in her own time before making up her mind. The doctor may then answer any further questions.

Follow-up

DMPA: 12-weekly/(150 mg) or 16-weekly/(250 mg) visit intervals.
NET OEN: 8-weekly/visit intervals.

Access to medical advice by telephone or in person must be easily available if anxieties arise between visits.

Complications and their management

1. Menstrual disturbances

a. The patient should be examined to exclude gynaecological causes of bleeding, especially pelvic inflammatory disease, or retained products of conception in postpartum women. Bleeding from a firm, well involuted, nontender uterus with a closed os is almost certainly not associated with significant pathology.

b. Pregnancy should be eliminated as a cause of amenorrhoea.

c. Treatment of nonacceptable bleeding:

(i) Oestrogenic combined pill, e.g. ethinyloestradiol 50 μg + lynestrenol 2.5 mg (Minilyn): 1 tablet daily for 22 days.

(ii) Conjugated oestrogens (Premarin) 1.25 mg: 1 tablet daily for 21 days. If withdrawal bleeding persists for longer than 7 days, the course may be repeated.

Either may also be tried with a view to inducing bleeding in unacceptable amenorrhoea, once the possibility of pregnancy has been eliminated.

(iii) Progestogen. The subsequent injection may be given early, e.g. at 8 weeks instead of 12.

(iv) Haemostatics. Aminocaproic acid (Epsikapron) 12 or 24 g per day. Ethamsylate (Dicynene) 500 mg every 6 hours. Although there are no published results, these regimens are claimed to be effective if used for the duration of heavy bleeding as in the treatment of bleeding with IUDs.

(v) Iron. Does not prevent bleeding but may be needed to correct anaemia.

None of these treatments may be effective. In the UK prolonged irregular bleeding is unlikely to be tolerated for more than a year and it is the commonest reason for discontinuing the method. Postpartum women are especially likely to be troubled and the bleeding may be heavy. However, many of those who have problems can be tided over the initial difficulties by one of the above regimens and encouraged to look forward to an improvement as the number of injections increases.

2. Weight gain
   a. Dieting is possible with DMPA, but it is difficult.
   b. Diuretics are useless as there is no evidence of fluid retention.
3. Delayed return of fertility. In practice a woman should not be expected to conceive for at least 8 months from the time of the last injection, i.e. 4–5 months after expiry of the calculated contraceptive effect. In those who fail to conceive thereafter, ovulation can be successfully induced with human gonadotrophins or gonadotrophin-releasing hormone, under appropriate specialist supervision.

RISK OF CARCINOGENESIS

DMPA

Cervix. No evidence of increased risk of dysplasia or carcinoma in situ. *Reason for disquiet:* two uncontrolled trials in the United States showed a greater prevalence in users, *BUT*
   1. they were selected populations with other increased risk factors,
   2. the changes were detected within one year of onset of treatment, whereas almost all known carcinogens take considerably longer to produce an effect.

Breast. No evidence of increased risk in women. *Reason for disquiet:* increased incidence of benign and malignant breast nodules in beagle bitches treated with 17α-hydroxyprogesterone derivatives (including DMPA), *BUT*
   1. beagles are particularly susceptible to progestogenic stimuli and the resulting tumours are quite unlike human breast tumours,
   2. Both WHO and CSM have concluded that beagles are an inappropriate species in which to test contraceptive steroids.

Endometrium. No clinical evidence of increased risk in women. *Reason for disquiet:* 2 of 12 surviving rhesus monkeys given 50 times the human contraceptive dose in a 10-year trial developed endometrial carcinoma, *BUT*
   1. DMPA is used in treatment of human endometrial cancer,
   2. it is possible that the carcinomata arose from epithelial cell plaques that are not found in humans, an analogous situation to that of the beagles and their breast nodules,
   3. the animals were 'wild' caught and there is no background information about them,
   4. no lesions developed in monkeys treated with the human dose or with 10 times the human dose (USAID Report 1980)

Liver. Two cases of benign hepatic tumour have been reported in users in the USA, *BUT*

1. combined OC users have a known increased risk,
2. a general increase in incidence in nonusers in the USA has been reported.

NET OEN

Breast and pituitary tumours are caused in rats on which it has a strong oestrogenic effect which is not found in humans. Rats are so different that they are considered unsuitable for testing contraceptive steroids for use in humans.

COMPARISON OF RISKS/BENEFITS BETWEEN INJECTABLES, ORAL CONTRACEPTIVES AND INTRAUTERINE DEVICES

The evidence available at present is summarised in Table 6.2.

Continuation rates are between 50 and 80% at the end of 1 year falling to 40 to 60% in the second year. This is higher than among OC users but lower than with IUDs. Ill-informed adverse publicity about the risks is an important reason for discontinuing this method.

CONCLUSION

There are no good scientific reasons why DMPA, and probably NET OEN should not be available to women in the same way as combined oral contraceptives and IUDs. Now that a long-term licence has been granted, the information available to general practitioners, obstetricians, gynaecologists and family planning doctors will be greatly increased. This will reduce the medical mismanagement which currently bedevils the use of injectables.

Women should always be given a fair account of the pros and cons of any method of contraception they are offered. They should have subsequent access to informed advice if they become anxious, whether or not the cause of their worry is actually related to the method. Routine administration of DMPA in association with postpartum rubella immunisation without adequate explanation to the patient or information to her general practitioner has done incalculable harm.

The greatest practical disadvantage in the use of either injectable preparation is the frequent occurrence of irregular bleeding. WHO held a three-day conference on this aspect of DMPA alone. The exact

Table 6.2 Comparison with combined OC and IUD on present evidence

Effect	DMPA	OC	IUD
Death	Nil	From cardiovascular disease especially older heavy smokers	From midtrimester septic abortion
Morbidity	Unnecessary dilatation & curettage for bleeding	Myocardial infarction, cerebrovascular accident, gall bladder disease	Pelvic inflammatory disease Anaesthetic related risks associated with insertion and removal
Subsequent fertility	Delayed	Minimal delay	Permanent infertility associated with PID (rare)
Efficacy method-related user-related	Very high Independent of user for several months	Very high Daily dependence on user	High Independent of user for several years
Cycle control	Nonexistent	Excellent	Good to fair
Excessive and/or prolonged bleeding	Frequent especially postpartum	Nil	Not common with smaller devices
Amenorrhoea	Frequent especially with long-term use	Uncommon	Not related
Systemic disturbances	Slight but definite. Probably not clinically significant	Slightly greater. Changes in coagulation factors of clinical significance	Possible iron-deficiency anemia
Fetal damage	Very rarely transient enlargement of clitoris	Very rarely limb deformities	Miscarriage may occur if left in situ Possible prematurity

cause is poorly understood and we do not have a really effective means of treatment. It is hoped that a remedy will soon be found or, better still, that we shall be able to identify potential bleeders beforehand. Despite these problems, injectables are safe, reliable and effective contraceptives which may be more suitable for some individuals than any of the other methods available.

REFERENCES

Detailed references to most of the information contained in this chapter may be found in:

Ceulaer K, de Gruber C, Hayes R, Serjeant G R 1982 Medroxyprogesterone acetate and homozygous sickle-cell disease. Lancet ii: 229–231

Department of Medical and Public Affairs 1975 Injectables and implants. Population Report Series K, no 1, March. (The George Washington University Medical Centre, 2001 S Street, NW Washington DC20009)

Fraser I S, Weisberg E 1981 A comprehensive review of injectable contraception with special emphasis on depot medroxyprogesterone acetate. The Medical Journal of Australia Jan 24, 1 (1 Suppl): 3–19

Howard G, Blair M, Fotherby K, Howell R, Elder M G, Bye P 1980 Clinical experience with intramuscular norethisterone oenanthate as a contraceptive. Journal of Obstetrics and Gynaecology 1: 53–58

Schardein J L 1980 Congenital abnormalities and hormones during pregnancy: a clinical review. Teratology 22: 251–270

USAID Ad Hoc Consultative Panel (July 1980) Report on Depot Medroxyprogesterone Acetate.

Wilson E 1976 Use of long-acting depot progestogen in domiciliary family planning. British Medical Journal 2: 1435–1437

7

John R. Newton

Intrauterine contraceptive devices

Intrauterine contraceptive devices (IUDs) have been a realistic method of contraception for approximately 70 years. They have developed from the wishbone and collar stud pessaries of the 19th century and the Richter, Grafenberg and Ota rings of the early 20th century.

In the 1960s silastic devices first appeared, in a wide variety of flexible plastic shapes, with a 'memory', i.e. they returned to their original shape after straightening. They were followed in the early 1970s by copper IUDs which had copper wire wound round silastic frames. Second generation copper devices were developed in the late 1970s with sleeves of copper instead of copper wire to try to prolong the active life; different shapes to reduce the risk of expulsion; and silver cored copper wire to decrease the risk of fragmentation.

Around this time hormone-releasing medicated devices were also developed (containing progesterone or levonorgestrel).

TYPES OF IUD

Those in current use can be divided into three categories — inert or plastic, copper and medicated. All have one or two threads attached. While the colour of the thread is often helpful in identifying the type of device which a patient has in situ, it is important to remember that the colour of the threads of some devices has changed over the years and therefore cannot be relied on as positive identification.

Inert devices

Plain silastic devices have been made in many different shapes and sizes. Most contain barium sulphate to make them radio-opaque. The Lippes Loop is the only inert device still available in the United Kingdom. The Saf-T-coil is no longer marketed here.

Lippes Loop

The four sizes are shown in Figure 7.1. Size C is the one most commonly used.

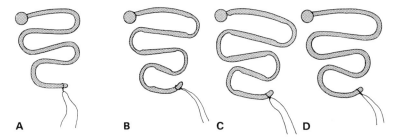

A   B   C   D

Fig. 7.1 Four sizes of Lippes Loop

Size	A	B	C	D*
Threads	2: blue	2: black	2: yellow	2: white
Length	26 mm	25 mm	27.5 mm	27.5 mm
Width	22.5 mm	27.5 mm	30 mm	30 mm
Composition	Polyethylene			
Duration of use	Indefinite			
Supplier	Ortho-Cilag Pharmaceutical Ltd.			

*Although the dimensions of sizes C and D are the same the latter is thicker and heavier.

Copper devices

Most devices now contain copper which increases contraceptive efficacy allowing the devices themselves to be smaller and easier to fit.

Gravigard (Copper 7: Cu 7)

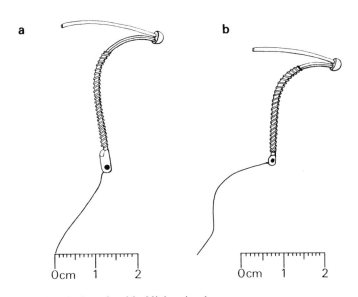

a
b

Fig. 7.2 a: Standard gravigard **b:** Minigravigard.

Size	Standard gravigard	Minigravigard
Threads	1: blue	1: blue
Length	36 mm	28 mm
Width	26 mm	22 mm
Composition	Polypropylene carrier with copper wire wound round the vertical stem	
	Surface area of copper = 200 mm²	
Duration of use	2 years	
Supplier	Gold Cross Pharmaceuticals	

Copper T (Gyne T)

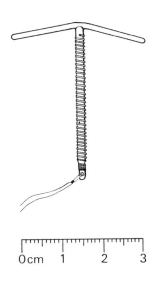

Fig. 7.3 Copper T (Gyne T)

Size	1 size only
Threads	2: white
Length	36 mm
Width	32 mm
Composition	Polyethylene carrier with copper wire wound round the upright stem. Surface area of copper = 200 mm^2 The 220C has 220 mm^2 of copper as 'sleeves'. The 380A has 380 mm^2 of copper, with 'sleeves' on the horizontal arm and wire on the stem.
Duration of use	3 years
Supplier	Ortho-Cilag Pharmaceutical Ltd

Multiload Cu 250

Available in standard and short sizes. A smaller multiload is being developed as are other varieties containing different amounts of copper.

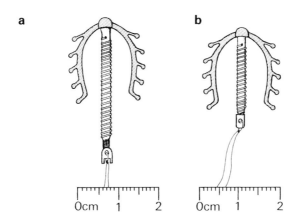

Fig. 7.4 a: Standard Multiload; **b:** Multiload short.

Size	Standard	Short
Tails	2: blue	2: blue
Length	36 mm	25 mm
Width	18 mm	18 mm
Composition	Polyethylene carrier with copper wire wound round the stem. The diameter of the copper wire used is slightly thicker than in other devices (0.3 mm in the standard and 0.4 mm in the short). Surface area of copper = 250 mm² in both sizes.	
Duration of use	3 years	
Supplier	Organon Laboratories Ltd	

Nova T: Novagard

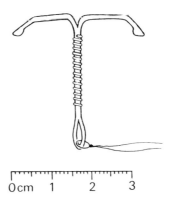

Fig. 7.5 Nova T: Novagard.

Threads	2: white
Length	32 mm
Width	32 mm
Composition	Polyethylene with copper wire which has a silver core wound onto the vertical double stem. Surface area of copper = 200 mm^2
Duration of use	5 years
Suppliers	Schering Chemicals Ltd Gold Cross Pharmaceuticals

Medicated devices

These are still being developed. The only one currently available is the Progestasert (Fig. 7.6). However, it is seldom inserted now because its use appeared to be associated with an unduly high incidence of ectopic pregnancy. Devices containing different doses of progesterone or synthetic progestogens are under trial.

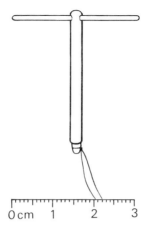

0 cm   1   2   3

Fig. 7.6 Progestasert

Threads	2: blue-black
Length	36 mm
Width	32 mm
Composition	Frame of ethylene/vinyl acetate with a reservoir of 38 mg progesterone which is delivered continuously into the endometrial cavity at a rate of 65 μg per day.
Duration of use	1 year
Supplier	Polcrone Ltd, Pharmaceutical Distributors, 11 Mount Road, Feltham, Middlesex.

MODE OF ACTION

The exact mechanism of action of IUDs is still unknown but may be summarised as follows:

1. All IUDs cause an inflammatory reaction in the endometrium so that phagocytic leucocytes may then engulf the sperm or fertilised ovum.

2. Copper devices alter the enzyme and trace element content of the endometrium. Copper is also spermicidal.

3. Steroid-releasing IUDs suppress the endometrium in a way similar to the progestogen-only pill. They also inhibit ovulation.

The IUD therefore interferes with the complex process of implantation and/or produces desynchronisation and endometrial changes that are not compatible with pregnancy.

EFFECTIVENESS

Contraceptive effectiveness varies from one device to another and in different studies, though when pregnancy, expulsion and removal rates are considered together no device appears to have a clear advantage. Maximum effectiveness is not dependent on either user action or motivation.

A pregnancy rate of 1–3 per 100 women years is frequently quoted (2–4 in the first year of use). The rates for pregnancy and removal for bleeding tend to fall slowly with time; the expulsion rate falls dramatically with time.

Data from the Oxford/FPA study (Vessey et al 1982) show:

1. Failure rates per 100 women years: Gravigard = 1.5; Lippes Loop C = 1.4.

2. A noticeable decline in failure rates with increasing age and duration of use.

INDICATIONS

1. Parous women who do not wish to take oral contraceptives or in whom their use is contraindicated.

2. Nulliparous women unable or unwilling to use another form of contraception.

3. Older women coming off the pill who do not wish to use another method or to be sterilised.

4. Women who believe that a heavy, regular period is essential for their continued good health.

5. The forgetful.

6. Women who see their partners sporadically.

CONTRAINDICATIONS

Absolute contraindications

1. *Known or suspected pregnancy.*

2. *When complete protection against pregnancy is essential.*

3. *Undiagnosed abnormal vaginal bleeding.* Once malignancy or other uterine pathology is excluded and menstruation returns to normal an IUD can be fitted.

4. *Suspected malignancy of the genital tract.* Appropriate gynaecological investigation should be instituted without delay. After local therapy for certain early lesions of the cervix (CIN II, carcinoma in situ) an IUD can be inserted.

5. *Active pelvic inflammatory disease (PID).* Fitting an IUD will increase the severity of infection.

6. *Copper allergy (rare)* and *Wilson's disease* (for copper devices only).

Relative contraindications

1. *Previous ectopic pregnancy.* IUD use compounds the risk of further ectopic pregnancy.

2. *Cervical or vaginal infection.* After successful treatment an IUD can be inserted. Simple erosions of the cervix are not a contraindication.

3. *Uterine fibroids.* Provided fibroids are not large and periods are not excessively heavy an IUD can be used. Yearly pelvic examination, to assess any change in the size of the fibroids is necessary.

4. *Abnormality of the uterine cavity.* The shape of the uterine cavity may be distorted and necessitate modification of the IUD before fitting, e.g. shortening of the transverse arm. With a double uterus 2 IUDs need to be fitted.

5. *A recent history of treated pelvic infection.* A single episode of confirmed PID adequately treated is not a contraindication to IUD use provided it has not occurred within the previous six months. However, several attacks, confirmed or otherwise, must be viewed with suspicion and alternative methods of contraception advised.

6. *Uterine scars* from previous surgery other than caesarean section.

7. *Menorrhagia.*

8. *Valvular heart disease.* There is a risk of subacute bacterial endocarditis. If a decision is made to insert an IUD, appropriate antibiotic cover should be given to patients with a prosthetic valve (Working Party of the British Society for Antimicrobial Chemotherapy 1982).

9. *Nulliparity.* This remains controversial. The increased risk of PID and possible impairment of future fertility has to be weighed against the inability or unwillingness of the woman to use another method.

10. *Insulin-dependent diabetes.* There is evidence that IUDs are associated with a higher failure rate in such women (Gosden et al 1982) but this is disputed.

11. *Lactation.* Evidence is accumulating that the risk of uterine perforation at insertion is greater in women who are breast feeding. This does not mean that IUDs should not be inserted but it does

indicate the need for particular care with insertion (Heartwell & Schlesselman 1983).

12. *Systemic corticosteroid treatment.*

ADVANTAGES

The main advantages are:

1. Repeated action by the user either daily or at the time of coitus is not required.

2. Visits for medical check-ups are infrequent.

3. With most IUDs more than 70% of users have few or no side effects.

4. Devices can be left in place for years — inert devices until menstruation ceases, copper devices for 2–5 years.

5. Once inserted, little patient compliance is required.

DISADVANTAGES

Minor problems

1. *Menstrual irregularities.* For the first few months intermenstrual bleeding or spotting may occur. This gets less with time. Premenstrual spotting for 2–3 days is also common.

2. *Heavy periods.* All copper and inert devices cause an increase in the amount of bleeding and/or the length of the period. Although there is great patient variability, the average loss in a normal cycle is 35 ml, in a copper IUD cycle 50–60 ml, and with inert devices 70–80 ml.

3. *Discharge.* Watery or mucoid discharge is common in women wearing an IUD, particularly a copper device. It is usually only of nuisance value. However, in the presence of low grade anaerobic infection the discharge may become foetid and unpleasant.

4. *Lower abdominal pain.* This may occur at insertion, particularly in nulliparous women, and last for a few days. Cramps may also accompany periods for the first few months and, if severe, may be an indication for IUD removal.

Major problems

See 'Complications and their Management' (p. 143)

CLINICAL MANAGEMENT

Assessment

Suitability for an IUD is based on history and physical examination.

1. *History.*

Take a history particularly in relation to the need to avoid pregnancy, past use of contraceptives, the acceptability of the method to the patient, and those conditions which might contraindicate its use. Only after a full discussion of the possible risks and benefits should an IUD be fitted.

2. *Examination*

a. Perform a pelvic examination to determine the size, shape and position of the uterus and to confirm that the appendages are normal.

b. Examine the vagina and cervix with a speculum to exclude abnormality and infection.

c. Take a cervical smear and examine the breasts if indicated by clinic or practice screening policy for women in the age group (Ch. 15).

d. Take the blood pressure if the woman is over 40 and has not had a normal blood pressure recorded in the previous 3 years (Ch. 15).

Choice of IUD

There are no hard and fast rules for selecting the right IUD for an individual patient. However, the following guidelines may aid or influence the choice.

1. In general copper IUDs are better than inert IUDs as they have a lower overall incidence of side effects.

2. Smaller devices, e.g. Minigravigard or Nova T are better suited to the small nulliparous uterus.

3. Larger devices, e.g. Copper T or Multiload are better suited to the larger parous uterus.

4. The patient's previous experience with IUDs may determine the choice, e.g. pregnancy with an inert device — try a copper IUD; pregnancy with a copper 7 — try a copper T or Multiload; unexplained expulsion of a small device — try a larger one (and vice versa).

5. Problems with a current IUD may dictate a change, e.g. menorrhagia or pain — change to a smaller device.

6. An inert device may be better in women who cannot be relied on to return for check-ups or periodic IUD exchanges.

Insertion

Instruments needed for insertion

Light source
Bivalve speculum
Bacteriological swabs (when appropriate)
Kidney dish (receptacle for insertion instruments)
10-inch sponge-holding forceps
Pair of scissors, uterine Sims curved on flat, 8-inch blunt ended
Malleable uterine sound graduated in centimetres
12-inch tissue forceps or single-tooth tenaculum with blunted tips.

Timing of insertion

1. Insertion during or shortly after a menstrual period is recommended as pregnancy is unlikely; the cervix is softer and the os open making insertion easier; postinsertion discomfort is less and bleeding is not noticed. However, an IUD may be inserted at any time during the menstrual cycle.

2. Immediately after delivery, termination of pregnancy or miscarriage. Expulsion rates are higher with immediate postpartum and postabortal insertion after the first trimester.

3. If not inserted immediately wait for two weeks after TOP or miscarriage and four to six weeks after delivery.

4. After unprotected intercourse as a postcoital contraceptive (Ch. 11).

General advice about technique

As the method of insertion is different for each device, insertion is safest if the manufacturer's instructions are followed meticulously. (P153–155)

1. Throughout the whole procedure a no-touch technique is employed so that only clean gloves need be worn. However, if manipulation of the sterile IUD in its holder is needed then sterile gloves must be used.

2. Following bimanual examination of the pelvis the cervix is exposed with a speculum while the patient lies in a modified lithotomy position. The left lateral position may be preferred.

3. The cervix is cleansed with antiseptic solution and grasped with 12-inch atraumatic forceps. This stabilises it and allows controlled IUD insertion while helping to achieve correct fundal placement.

4. A fine uterine sound is gently passed to determine the depth and direction of the uterine cavity and the direction and patency of the cervical canal.

5. The device is loaded into the introducer in such a way that it will lie flat in the transverse plane of the uterine cavity when it is released.

6. The device should not remain in the introducer tube for more than a few minutes lest it lose its shape.

7. The introducer tube is carefully inserted through the cervical canal, the IUD released according to the specific instructions for each device and the introducer withdrawn.

8. Following insertion further sounding of the canal to exclude the low lying IUD is necessary. Good fundal placement is essential to achieve a low incidence of expulsion and pregnancy.

9. The thread(s) of the tail of the IUD should be trimmed with long scissors to about *3 cm from the external os.*

NB (i) Some experienced IUD fitters do not use forceps to stabilise the cervix or use a sound to determine the size of the uterine cavity. Though it does seem likely that their technique is less traumatic for the patient it is not known whether it leads to better or worse results than those achieved by IUDs fitted in the recommended manner.

(ii) The skill and attitude of the fitter may influence the incidence of unwanted effects more than the type of device fitted.

(iii) Doctors who wish to insert IUDs are recommended to undergo the appropriate training (Ch. 1).

Problems at insertion

Pain: vasovagal syncope. Severe pain associated with vasovagal syncope — cervical shock — is rare and no deaths have been reported. It is usually caused by distension of the internal cervical os with the sound or introducer. In severe cases the insertion procedure should be stopped, the patient placed head down and the airway maintained. However, the majority of cases are mild and after an interval the operator can continue, using 1% local analgesic gel or a local analgesic injection applied to the cervical canal.

If shock occurs as the IUD insertion is being completed the IUD can be left in place and the patient allowed to recover.

The incidence of cervical shock may be reduced by inserting the IUD during menstruation or, in the anxious patient, by applying 1% local analgesic gel to the cervical canal. All 'cervical shock' reactions have become less common since smaller devices and narrower introducer tubes have become available.

Epileptic attacks are sometimes precipitated by IUD insertion. They are not necessarily associated with pain and usually occur two to three minutes after the stimulus.

Perforation. Occurs in about 1 in 1000 insertions. It may cause sudden pain or bleeding but often goes unrecognised. If suspected, immediate gynaecological opinion is required. Devices can often be removed from the peritoneal cavity by laparoscopy though sometimes laparotomy may be necessary.

Failure to insert. This may be due to anatomical abnormality, patient anxiety or poor operator technique. If unusual difficulty is encountered the fitting should be abandoned and the patient asked to return at a later date to see a more experienced doctor or referred to a gynaecologist for insertion possibly under general anaesthesia.

Resuscitation measures in IUD clinics

1. IUD insertion should always be carried out in a calm, relaxed, unhurried atmosphere and with a gentle reassuring approach to the patient at all times.

2. At the earliest sign of vasovagal attack insertion may need to be abandoned or the device removed. The patient should be kept supine, the head lowered and the legs elevated to the vertical position if necessary in order to restore blood flow.

3. A clear airway must be maintained — probably no more than supporting the chin will be necessary. Any tight clothing, especially round the neck, should be loosened.

4. A Brook Airway size 900 should always be at hand.

5. Overtreatment should be avoided but simple procedures instigated at the first sign of trouble.

6. Correct positioning should always take precedence over more heroic procedures.

7. Where a persistent bradycardia is in evidence, atropine may be given intravenously in a dose of 0.6 mg. It should be borne in mind always that the injudicious use of drugs might do more harm than good.

8. If oxygen is available it may be administered using an Ambu bag.

9. In the very rare situation where the patient fails to regain consciousness she should be transferred by ambulance to an accident/emergency department or the nearest intensive care unit.

Instructions to patients

These are not complicated and refer to checking that the IUD is in place and to coping with side-effects.

1. *Feeling for the IUD thread(s).* Instruct the woman how to feel for the thread(s) coming through the cervical os. This can cause problems since the thread(s) of modern devices are soft, not easily felt and the woman becomes anxious when she fails to locate them.

It is much more important to instruct her to feel the cervix each month after the end of her period and to ensure that no firm plastic is protruding. If she can feel the plastic end of the device it means that it is partially expelled and she should consult her doctor.

2. *Bleeding after insertion.* This is of no consequence in the first 24 hours and reassurance is all that is required.

3. *Pain after insertion.* This is similar to menstrual cramps and can be controlled by simple analgesics, e.g. paracetamol or aspirin.

4. *Additional contraception.* This is not needed.

5. *Intercurrent therapy.* Medical treatment, e.g. with antibiotics, has no adverse effect on IUDs. It has been suggested that patients with copper IUDs should not have pelvic short-wave diathermy, which might induce heating in the copper coil. The risk is theoretical. Penotrane pessaries should not be used by women who have a copper IUD. Reports that the use of antiinflammatory drugs such as corticosteroids and aspirin increase the failure rate with IUDs have still to be confirmed (Buhler & Papiernik 1983).

Follow-up

1. The patient should be seen 6–8 weeks after insertion and annually thereafter unless symptoms develop necessitating an earlier appointment.

2. At each visit she should be asked about her menstrual pattern, pelvic pain and vaginal discharge.

3. Pelvic examination should be carried out to exclude any abnormality, to confirm that the IUD is in position and to take a cervical smear if screening guidelines dictate.

COMPLICATIONS AND THEIR MANAGEMENT

Minor complications are usually little more than a nuisance although if they persist they may lead to a request for removal of the device. Major complications require careful assessment and management.

Abnormal bleeding

See Table 7.1.

Table 7.1 Abnormal bleeding associated with IUDs

Symptoms	Features	Management
Heavy or prolonged periods	Common complication especially with larger devices. Heavy periods many months after insertion may be due to movement of the IUD in the uterine cavity, partial expulsion or gynaecological pathology. Prolonged bleeding and discomfort may indicate pelvic infection	With sympathetic reassurance many patients will tolerate such bleeding and keep the IUD. Mefenamic acid (Ponstan) or antifibrinolytic agents may decrease blood loss but if heavy bleeding persists the device should be removed and replaced with a smaller copper type or alternative form of contraception. Heavy bleeding as a new symptom demands vaginal examination and probably removal of the IUD; observation of subsequent period(s); gynaecological referral if necessary. If normal menstrual cycles return another device may be inserted
Intermenstrual bleeding (IMB)	Especially likely in the luteal phase	This should be disregarded during the first few months after insertion. If it develops months or years later in an otherwise trouble-free patient remove the IUD. If it persists refer for gynaecological investigation
Intermenstrual spotting (IMS)	May occur at any time in the cycle but most commonly 1–2 days before a period	Warn patients to expect this and reassure them. If troublesome treat with a nonsteroidal anti-inflammatory agent (NSAI). If persistent examine the cervix to check for disease and partial expulsion of the device

Table 7.1 (cont.)

Symptoms	Features	Management
Amenorrhoea	Any patient presenting with a missed period or more prolonged amenorrhoea should have pregnancy excluded by clinical examination and/or a pregnancy test. NB The pregnancy may be ectopic	If pregnant (p. 147). If not pregnant and the uterus is of normal size reassure her that her cycle will probably soon return to normal. If amenorrhoea persists for 6 months, appropriate investigation is indicated
Postcoital bleeding (PCB)	Rarely, if ever, caused by an IUD	Examine the cervix, take a smear and refer for gynaecological opinion if necessary

Pain

See Table 7.2.

Table 7.2 Pain associated with IUDs

Symptoms	Features	Management
Dysmenorrhoea	Similar to spasmodic dysmenorrhoea	Simple analgesics may be tried and if unsuccessful replaced by NSAI for the first 2–3 days of each period. Unrelieved symptoms may require IUD removal.
Dyspareunia	Although IUDs never cause superficial dyspareunia they are occasionally associated with deep dyspareunia and may sometimes cause the male partner to suffer a sharp pain during intercourse	Any underlying cause should be treated but if no cause is found the IUD should be removed and the effect of this assessed. When the male partner complains of pain during intercourse the patient should be examined to exclude a partially expelled device or a sharp thread protruding from the os. In both cases the IUD should be removed and replaced, or a different IUD may be tried later
Lower abdominal pain	Dull ache not related to periods. Often no cause is found	Exclude pelvic infection (p. 146) and partial expulsion of the IUD. If it persists and no cause is found the device may have to be removed

Pelvic infection

The exact prevalence and incidence of pelvic inflammatory disease (PID) associated with IUDs are difficult to establish as the diagnosis is usually based on clinical judgement and there are no standard diagnostic criteria. It is common for patients who have some vaginal discharge and/or uterine pain to be labelled as suffering from PID, yet these two symptoms do not by themselves signify infection.

However PID is more common in IUD users especially among young nulliparous women (Kaufman et al 1983). As it can cause subsequent infertility, the risks should always be fully discussed before the IUD is inserted. The risk is compounded if the woman has many sexual partners because of her increased risk of catching a sexually transmitted disease.

Pelvic actinomycosis has been reported in women with inert IUDs (Duguid 1983) but is rare. Actinomyces-like organisms are often identified on a Papanicolaou smear but if the patient is symptom-free the IUD may be left in situ and no treatment is required.

Diagnosis

1. Clinical features can range from virtually none to those of acute sepsis. Acute PID, unilateral or bilateral, is characterised by fever in excess of 38°C, uterine and pelvic tenderness, palpable pelvic adnexal mass, foul-smelling vaginal discharge, elevated white blood count and erythrocyte sedimentation rate.

2. High vaginal and cervical swab culture *may* identify aerobic or anaerobic organisms or both.

3. Laparoscopy can be used to confirm the diagnosis in doubtful cases.

Management

In mild cases the diagnosis should be established and treatment with antibiotics started. If there is no response after 24 hours the IUD should be removed.

In moderate cases with more definite clinical signs the IUD should be removed before starting antibiotic therapy and the patient referred for a gynaecological opinion.

Severe cases associated with marked lower abdominal pain and fever should be admitted to hospital.

Any patient with a past history of PID and who is suspected of developing pelvic infection should be treated. Most cases are mild and respond successfully to antibiotics such as metronidazole (Flagyl) amoxycillin (Amoxil) or tetracycline for a full seven days.

If *Actinomyces israelii* is identified on culture the device should be removed and treatment with penicillin instituted.

Pregnancy

Intrauterine pregnancy

The pregnancy may go successfully to term. However, the incidence of spontaneous septic midtrimester abortion, bleeding, amnionitis, premature labour and delivery and perinatal mortality are all increased if the IUD is left in place. There is no evidence that the presence of an IUD during pregnancy increases the risk of congenital abnormality.

Management. Up to 12 weeks gestation, if the threads are visible and the device moves easily, remove it gently.

Beyond 12 weeks, or if no threads are visible, or if the device does not move easily, leave it in situ. Refer early for antenatal care. Advise the patient to seek immediate medical attention if she develops bleeding, fever, amniotic fluid leak, etc.

If left in situ throughout pregnancy the device is usually expelled before or with placenta and membranes. If not, X-ray or ultrasound should be used early in the puerperium to locate it.

Extrauterine or ectopic pregnancy

Ectopic pregnancy occurs in 1 in every 27 pregnancies resulting from IUD failure, compared to 1 in 200 among women not using contraception. IUDs may protect more effectively against intrauterine rather than extrauterine pregnancy or IUD-related pelvic infection may cause tubal damage which predisposes to ectopic implantation.

Management. Early diagnosis is vital and any suspicion of ectopic pregnancy requires an immediate gynaecological opinion.

Classical features are amenorrhoea for 6–8 weeks accompanied by symptoms of pregnancy, severe lower abdominal pain with or without vaginal bleeding which is often scanty and black but these are not always present. The period may not even be late. The diagnosis should be suspected if there is *any* unexplained pelvic pain or lower abdominal cramps or *any* irregular bleeding, especially if a period is scanty, late or missed.

Vaginal discharge

If profuse or persistent, cervical and vaginal swabs should be taken to exclude infection. If no infecting organism is identified or if treatment is not successful then, as a last resort, replacement with a similar IUD but with the threads cut off may cure the problem. Subsequent removal of this IUD will obviously be more difficult.

LOST THREADS AND LOST DEVICES

Threads which are too long can be a nuisance to the patient. They can cause irritation and if pulled on can partially remove the device. Threads which are too short can lead to difficulty in finding the IUD and, unless they have become visible after the next menstrual period, become numbered among 'lost' threads. To avoid both these problems threads should be cut, at the time of insertion, 3 cm from the external os.

Lost threads

When threads are not visible at the external os, the device may be in the uterine cavity, embedded in the uterine wall, in the peritoneal cavity, or expelled altogether from the body.

To look for threads in the clinic or surgery:

1. Carefully expose the cervix, in a good light, as this will allow short threads in the cervical canal to be seen.

2. If no threads are seen, apply an Allis tissue forceps to the cervix and gently sound the uterus. Any device in the canal or uterus will usually be felt by the sound. A 4 mm Karman catheter attached to a 20 ml syringe can then be passed into the uterus, suction applied via the syringe and the catheter gently withdrawn. This will often bring down the threads.

3. If this fails, gentle exploration of the uterine cavity with an IUD removal hook may bring down the threads. This instrument may be a metal hook, a plastic spiral (Mi-Mark Helix) or a notched plastic sound (Emmett Thread Retriever). However, this procedure may displace the IUD and if it does, the device should be removed and a new one fitted.

4. If these procedures fail the IUD should be presumed lost and the patient referred for ultrasonography (p. 150).

Expulsion of an IUD

This is a problem associated with all devices. The risk is greatest in the first three months after insertion when 50% of all expulsions occur. The patient is often unaware of what has happened.

The skill of the fitter in ensuring correct fundal placement is an important factor in reducing expulsion rates.

Complete expulsion

Although expulsion of the whole device into the vagina is uncommon patients should be told to examine towels and tampons, especially in the first three months, to check this. They should also report unexplained pelvic pain or intermenstrual bleeding as these may signify expulsion.

If the patient wishes, another IUD may be inserted. Second insertions, even of the same type of IUD, are associated with lower expulsion rates than first insertion. Alternatively a T-shaped device or a Multiload, both of which have low expulsion rates, may be tried.

Partial expulsion

This is much more common than complete expulsion. Part of the device, usually the end of the vertical stem, is found protruding from the cervical canal. The patient or doctor may notice this or it may present as unexplained pain or intermenstrual bleeding. The device should be removed, and if the patient wishes, another inserted.

The lost IUD

Lost inert devices have remained in the peritoneal cavity for years without producing problems. However, they do tend to migrate in time and may be found anywhere in the abdomen. Linear devices theoretically cause few problems, while closed devices, like the Antigon triangle, can cause intestinal obstruction.

Copper devices seem to cause a sterile inflammatory reaction and rapidly become adherent to the omentum or bowel.

Abdominal pain may be present, though prior to this the patient may be otherwise asymptomatic.

All devices should be removed from the peritoneal cavity as soon as possible, although some authorities believe there is less urgency to remove linear inert devices in asymptomatic patients. That decision, however, will be made by the gynaecologist to whom the patient is referred.

The flow diagram shows the management sequence:

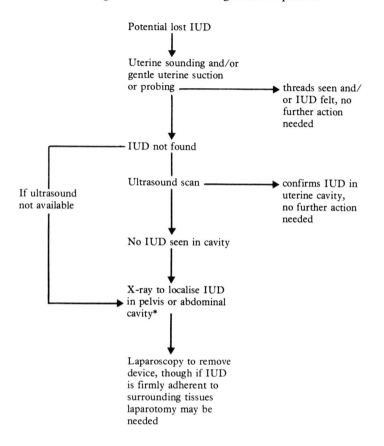

Flow diagram: Lost IUDs

Displaced IUD within the uterine cavity

Occasionally rotation of the IUD within the uterine cavity can occur. This will cause pain and bleeding and will necessitate removal of the device. Rarely downward displacement may lead to cervical perforation.

*Since there is no scientific evidence that the delivery of ionising radiation in the first 2 weeks of gestation in humans is teratogenic the need to observe the 10-day rule is being questioned.

In both cases referral to a gynaecologist for removal is advised as general anaesthesia may be necessary.

REMOVAL

An IUD may be removed to allow a patient to become pregnant, because contraception is no longer needed, or because a change of contraceptive method is desired. Complications, especially bleeding and pain, pregnancy or expulsion may also necessitate removal. Compared with smaller copper devices, larger inert devices are associated with fewer pregnancies and expulsions but are more frequently removed for bleeding and pain.

Unless a pregnancy is desired IUDs should not be removed in the latter half of the menstrual cycle if intercourse has occurred within the preceding seven days.

Manufacturers recommend that copper bearing devices be replaced every 2–5 years. The Gravigard should be replaced after 2–3 years, the Copper T after 3 years, the Multiload Cu250 after 3 years and the Novagard and Nova T after 5 years. However, recent data do not show an increase in pregnancy rates when they are left in place for longer.

The Progestasert should be replaced after 1 year.

At the menopause the IUD should be removed 12 months after the last menstrual period.

Technique

Visible threads

This is most easily carried out during a period but may be undertaken at any time in the cycle provided continuing contraceptive cover is *not* required (see above).

1. Check the size and position of the uterus by bimanual examination.
2. Expose the cervix with a speculum.
3. Grasp the retrieval thread(s) firmly near the cervical canal with a pair of straight artery forceps.
4. Apply gentle downward traction.

Usually the IUD will be withdrawn without difficulty and with minimal pain. If resistance to removal is encountered or if the patient experiences pain stop traction and then:

1. Grasp the cervix with tissue forceps and apply gentle traction to straighten out the cervical canal and uterine cavity.

2. Continue traction on the IUD threads and remove in the usual way.

3. Sometimes it may be necessary to dilate the cervical canal to 3 mm (Hegar 3).

4. It is rarely necessary to administer local or general anaesthesia.

Threads that break

If, during removal, a thread or threads break, the stem of the IUD is likely to be in the cervical canal. It may be removed with tissue forceps or a 'hook' but only experienced doctors should attempt to do this in a clinic or practice situation.

1. Apply tissue forceps to stabilise the cervix.

2. If necessary, use a local analgesic gel applied to the cervical canal or an intracervical injection of analgesic.

3. Insert straight artery forceps into the canal and grasp and lower part of the IUD.

4. Remove the IUD with gentle traction.

5. If this fails a small IUD removal hook can be passed into the uterine cavity and locked round the IUD, which is then removed with gentle traction.

6. If all these procedures fail, proceed as for lost IUD (p. 149).

Threads not visible (p. 148)

CONCLUSION

For about 8% of women in the UK the IUD is currently providing safe, effective, acceptable and relatively trouble-free contraception.

The risks of pelvic infection and ectopic pregnancy, in particular, make it an unsuitable first choice for some, while for others the minor side effects may lead to dissatisfaction with the method.

Careful selection of patients coupled with expert fitting and understanding follow-up contribute greatly to successful long-term use.

REFERENCES

Buhler M, Papiernik E 1983 Successive pregnancies in women fitted with intrauterine devices who take anti-inflammatory drugs. Lancet i: 483

Duguid H 1983 Actinomycosis and IUDs. International Planned Parenthood Federation Medical Bulletin, vol 17, no 3

Gosden C, Steel J, Ross A, Stringbett A 1982 Intrauterine contraceptive devices in diabetic women. Lancet i: 530–535

Heartwell S F, Schlesselman S 1983 Risk of uterine perforation among users of intrauterine devices. Obstetrics and Gynecology 61: 31–36

Kaufman D W, Watson J, Rosenberg L et al 1983 The effect of different types of intrauterine devices on the risk of pelvic inflammatory disease. Journal of the American Medical Association 250 (no 6): 759–762

Vessey M, Lawless M, Yeates D 1982 Efficacy of different contraceptive methods. Lancet i: 841–842

Working-party of the British Society for Antimicrobial Chemotherapy 1982 Report on the antibiotic prophylaxis of infective endocarditis. Lancet ii: 1323–1326

Appendix

DEVICE-SPECIFIC ADVICE

Insertion of Lippes Loop

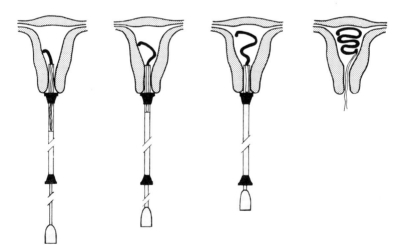

Fig. 7.7 Insertion of Lippes Loop

a. Load the loop into the introducer by pulling back on the plunger.

b. Cut thread, replace the plunger in the introducer tube and insert into the uterine cavity until the flange comes up against the cervix.

c. Hold the forceps (applied to stabilise the cervix) and introducer together and gently push the plunger inwards to expel the Lippes Loop into the uterine cavity.

d. Remove the introducer and plunger.

Insertion of Gravigard

a. Fold the device flat and slide it backwards into the introducer tube.

b. Move the blue cervical stop, which should lie in the same plane as the folded IUD, along the introducer tube to the correct depth of the uterine cavity already measured by sounding. Insert the introducer into the uterine cavity to the appropriate depth. Free the thread by pressing the thread-releasing clip from the rod.

c. Hold the forceps (as above) and plunger together withdraw the introducer tube 3 cm to free the horizontal arm of the device, and then gently push the IUD towards the fundus with the plunger so that the transverse arm lies in the widest part of the uterine cavity.

d. Remove the introducer tube and plunger.

Insertion of Copper T

a. Bend the horizontal arms downwards and insert into the introducer tube alongside the vertical arm. An arm locker has been developed to assist loading.

b. Set the cervical flange as for the Gravigard and insert the introducer tube into the uterine cavity while exerting gentle traction with the forceps.

c. When the tip of the introducer reaches the fundus — the cervical flange should at this point be up against the cervix and in the horizontal plane — retract the introducer tube 1–2 cm while holding the plunger stationary. This releases the arms of the Copper T.

d. Withdraw the plunger while holding the inserter tube stationary so as to leave the Copper T at the fundus.

e. Withdraw the inserter tube slowly.

Insertion of Multiload

Unlike other IUDs the Multiload has no plunger and no method of retaining the IUD in the inserter tube. It appears to be the easiest IUD to insert.

a. Apply forceps to the anterior lip of the cervix and sound the uterus.

b. Set the opaque white cervical flange at the appropriate point on the introducer tube.

c. Apply gentle traction on the forceps and use the introducer to insert the Multiload through the cervical canal into the uterine cavity until fundal resistance is felt and the cervical flange comes up against the cervix.

d. Gently withdraw the introducer using a pill-rolling movement to free the tube without dislodging the IUD.

Insertion of Nova T, (Novagard)

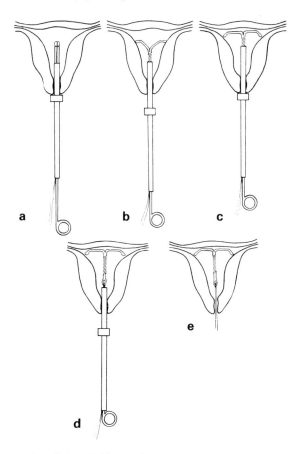

Fig. 7.8 Insertion of Nova T, Novagard.

a. Sound the uterus, set the cervical stop and load the device into the introducer by pulling on the threads.

b. Holding the threads taut, insert the plunger into the introducer, pass it into the uterine cavity as before and release the transverse arm by pulling back on the introducer tube (holding the plunger and forceps steady) until it reaches the ridged portion of the solid rod. A distinct click will be felt as the arms are released.

c. Push the introducer upwards until the cervical stop reaches the cervix to ensure correct fundal placing.

d. Release the IUD by pulling the introducer tube downwards until the plunger is completely inside it.

e. Remove the plunger and then the introducer tube with a pill-rolling action.

Barrier methods

Occlusive pessaries (caps)
Diaphragm
Cervical cap
Vault cap
Vimule
Clinical management of the patient
  Diaphragm
  Cervical cap
  Vault cap and vimule
  Schedule of visits

New barriers

Home made barriers

Condoms
Types

Spermicides
Creams and jellies
Vaginal suppositories
Foaming tablets
Aerosol foam
C-film

Vaginal contraceptive sponge

Coitus interruptus
History
Description

Coitus reservatus

Coitus interfemora

Coitus designed to avoid the vagina

Risks/benefits

Barrier methods aim to achieve contraception by preventing access of live spermatozoa to the ovum. They include the oldest human attempts at fertility control and comprise:

1. Occlusive pessaries for use by the female.
2. Condoms for use by the male.
3. Spermicides.
4. Adaptations of coital technique.

Increase in their use (Bone 1980) reflects a response to adverse publicity about the risks of hormonal contraception and IUDs. For optimum efficacy barrier methods rely on clear understanding and teaching of how they should be used in relation to individual needs. The adviser can help minimise the loss of spontaneity and pleasure which may be associated with any method whose use is directly related to the time of coitus.

OCCLUSIVE PESSARIES (Caps)

For thousands of years women have attempted to avoid pregnancy by blocking the access of sperm to the cervix. Various materials such as sponges and pads of cotton were used but occlusive pessaries did not appear until the 19th century. The German gynaecologist Hasse (using the pseudonym Mensinga) is credited with introducing the diaphragm in 1882. Wilde described the use of a rubber check pessary made to a wax model of the patient's cervix almost 50 years earlier. This type of cervical cap was used in Britain before the diaphragm but by the 1950s the diaphragm or Dutch cap represented the main choice of women seeking to control their own fertility.

Types

Four occlusive pessaries are in current use — the diaphragm, the cervical cap, the vault cap and the vimule.

Diaphragm

This is the most commonly used occlusive cap. It consists of a thin latex rubber hemisphere, the rim of which is reinforced by a flexible flat or coiled metal spring. Sizes, measuring the external diameter of the rim, range from 50–100 mm in steps of 5 mm, 70–80 mm being the sizes most commonly used (Fig. 8.1). Availability of intermediate, half

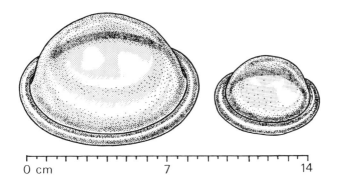

Fig. 8.1 Large and small diaphragms.

sizes (i.e. 2.5 mm increments in the flat spring type) and those at the extremes of the size range is decreasing. In clinical practice the reduced tension in the rim of the coil-spring will usually overcome the lack of a 'half size' in the flat-spring type for most patients.

Variations

1. The flat-spring diaphragm has a firm watch spring (Fig. 8.2), is easily fitted and remains in the horizontal plane on compression. It is suitable for the normal vagina and is often tried first.

2. The coil-spring diaphragm has a spiral coiled spring (Fig. 8.3), considerably softer than the flat-spring, and is particularly suitable for a patient who has strong muscles or who is sensitive to the pressure of the flat-spring type. With the largest sizes, handling may be slightly less easy because of a tendency to twist, especially with the non-opaque, light brown rubber variety.

Fig. 8.2 Flat-spring Durex diaphragm to show rim in cross-section.
Fig. 8.3 Coil-spring Ortho diaphragm to show rim in cross-section.

3. The arcing diaphragm (Fig. 8.4) combines features of both the above and exerts strong pressure on the vaginal walls. It is available on special order and it may be particularly useful when muscles have poor tone or the length of the cervix makes fitting of the more common types of diaphragm difficult.

Mode of action

The diaphragm lies diagonally across the cervix (Fig. 8.11), the vault and most of the anterior vaginal wall and acts as a barrier between the cervix and penis during intercourse. Choice of the correct size is important both to ensure that the diaphragm will not be displaced during intercourse and that the external os will not be exposed to the main volume of ejaculate. Since the vagina is known to alter in size during intercourse exact fitting can never be achieved.

Effectiveness

Because a sperm-tight seal between the rim of the cap and the vaginal walls is quite impossible, the use of a spermicide in conjunction with all caps is advocated in order to provide maximum effectiveness (p. 181). However, if the use of a spermicide is a crucial factor in making the method unacceptable to the patient it may be wise to concede its use and depend on the barrier function of the cap alone. It is then important to explain to the patient that the method may not be quite so reliable as when a spermicide is used as well.

Pregnancy rates vary widely, but with proper use are less than 5 per 100 women years.

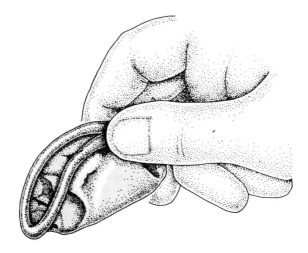

Fig. 8.4 Arcing diaphragm.

A comprehensive study (Vessey & Wiggins 1974) gives insight into important variable factors and shows that the diaphragm is a reliable and acceptable contraceptive, for the highly motivated couple. The results are summarised in Table 8.1. It should be noted, however, that these figures apply only to women over 25 years of age who had used the diaphragm for 4 months before entering the study and who had been carefully instructed in its use. They do indicate, however, how effective the method can be for well-motivated women who have learned to use it correctly.

Causes of failure, apart from poor motivation, are incorrect insertion or fitting, and displacement or defect in the cap due to careless handling.

Table 8.1 Standardised use-effectiveness pregnancy rates for the diaphragm according to certain characteristics of the subjects at entry to the study (Vessey & Wiggins 1974)

Characteristic	Standardised pregnancy rate per 100 woman years at risk	Rates standardised for
Age (years)		
25–29	2.8	Parity
30–34	2.6	Completion of family
35–39	1.5	Duration of use
Parity (no. of previous births)		
0	1.5	Age
1 or more	2.6	Completion of family
		Duration of use
Completion of family		
Yes	2.1	Age
No or uncertain	2.9	Parity
		Duration of use
Duration of use of diaphragm (months)		
5–23	3.4	Age
24–59	2.8	Parity
60 or more	1.7	Completion of family

Indications

1. Choice by the patient of a barrier method.

2. As an alternative to hormonal contraception for a woman who should not or does not wish to take oestrogen or progestogen.

3. Need for intermittent or infrequent yet predictable contraception.

Contraindications

1. Deficiency of vaginal tone, cystocoele, urethrocoele, etc.

2. Inadequate retropubic ridge.

3. Psychological aversion or inability to touch the genital area.

4. Inability to learn insertion technique.

5. Lack of hygiene or privacy for insertion, removal and care of the cap at home.

Advantages

1. No systemic side effects.

2. Reduction in the risk of carcinoma of the cervix (Wright et al 1978).

3. Reduction in the risk of sexually transmitted and pelvic inflammatory disease (Kelaghan et al 1982).

Disadvantages

1. A coitus-related method: involves preparation and loss of spontaneity.
2. Loss of cervical and some vaginal sensation.
3. May cause discomfort to the partner.
4. Must be fitted initially and checked regularly by a trained doctor or nurse.
5. Some women may experience frequency and dysuria both during and after use.

Cervical cap

This cap is shaped like a thimble and is designed to fit closely over the cervix. It is held in place by precise fitting onto the cervix and by suction, not by spring tension as in the diaphragm.

Variations

1. The cavity rim cap, made of firm pink rubber with an integral thickened upper rim incorporating a small groove intended to increase suction to the sides of the cervix, is the most commonly used (Fig. 8.5).

Fig. 8.5 Cervical cap.

Sizes, measured from the internal diameter of the upper rim, range from 22–31 mm in 3 mm stages.
2. A soft, brown rubber check pessary which is lighter than the cavity rim cap.

3. A firm, polythene cap shaped like a plain cap and in one size only is still under trial.

4. Intracervical caps and stem pessaries are not recommended.

Mode of action

By covering the cervix, the cap acts as a physical barrier to the entrance of sperm into the cervical canal.

Effectiveness

Accurate figures for reliability are not available.

Indications

The wish to use an occlusive pessary, in a woman who is unsuitable for any other type, provided the cervix is normal and healthy with parallel (not conical) sides pointing down the axis of the vagina and not acutely backwards.

Contraindications

1. Short, damaged, conical or unhealthy cervix.
2. Cervical discharge.
3. Inability to reach the cervix with the fingers.

Advantages

1. Suitable for patients with poor muscle tone and some cases of uterovaginal prolapse.
2. Not felt by the male partner.
3. No reduction of vaginal sensation.
4. Fitting unaffected by changes in the size of the vagina either during intercourse or as a result of changes in body weight.
5. Unlikely to produce urinary symptoms.

Disadvantages

1. Requires accurate selection of cap size and fitting to avoid displacement during intercourse.
2. Self-insertion and removal of the cervical cap are more difficult than with the diaphragm.

Vault cap

This cap, made of rubber, is an almost hemispherical bowl with a thinner dome through which the cervix can be palpated (Fig. 8.6). It is designed to fit into the vaginal vault, stays in place by suction and covers but does not fit closely to, the cervix. Five sizes are available ranging from 55 to 75 mm in 5 mm steps.

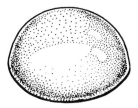

Fig. 8.6 Vault cap (Dumas).

Indications

    1. Wish to use an occlusive cap.
    2. Unsuitable for, or unable to use a diaphragm.
    3. Unsuitability of the cervix either because of its shape, position or condition for a well-fitting cervical cap.

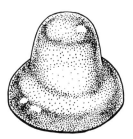

Fig. 8.7 Vimule.

Vimule

This is a variation of the vault cap with a thimble-shaped prolongation of the dome (Fig. 8.7). There are three sizes — small (45 mm), medium (48 mm) and large (51 mm).

Indications

It is used specifically for the patient requiring a vault cap to accommodate a cervix which is so long that it prevents suction being exerted by a Dumas on the vaginal vault.

Clinical management of the patient

Having ascertained by discussion that the woman wants to use an occlusive pessary and that it is socially and psychologically acceptable, an initial examination is made to note the following:

1. Position and condition of the uterus and cervix.
2. Length of vagina and muscle tone.
3. Type of retropubic ridge.
4. Measurement of the distance between the posterior fornix and the posterior aspect of the symphysis pubis.

Screening procedures should be carried out according to the routine clinic or practice policy.

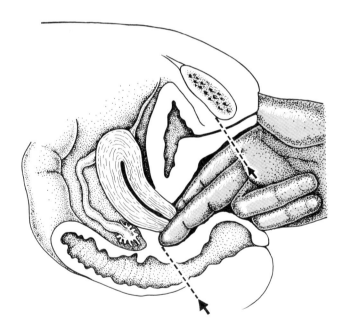

Fig. 8.8 Estimating the size of diaphragm to be fitted.

Diaphragm

Selection and fitting. 1. A diaphragm, corresponding roughly in size to the distance between the posterior fornix and the symphysis pubis, is chosen (Figs 8.8 and 8.9).

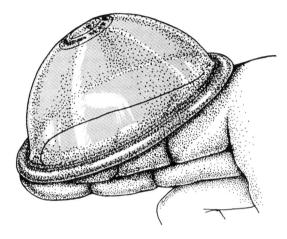

Fig. 8.9 Size of diaphragm on hand.

2. With the patient supine, the labia are separated, the diaphragm is compressed and inserted into the vagina, downwards and backwards to the posterior fornix, before being released (Fig. 8.10).

3. The anterior rim is tucked behind the pubic rami and the position of the cervix confirmed (Fig. 8.11).

4. Secure fitting is checked. When the patient strains down the anterior rim of the cap should not project or slip.

5. Too large a cap may project anteriorly, be immediately uncomfortable or become uncomfortable or distorted after wearing (Fig. 8.12a). This may be a cause of nonpersistence in its use.

6. If the cap is too small a gap will be felt between the anterior rim and the posterior surface of the symphysis pubis, or it may even be inserted in front of the cervix (Fig. 8.12b).

7. Persistent anterior protrusion of the diaphragm may be due to a mild cystocoele or a poor retropubic ridge and may be discovered only after the patient strains or stands up. In this case a cap other than a diaphragm is required.

8. The flat-spring cap is usually used dome upwards and if reversed (to increase retention behind the symphysis pubis) is slightly more

difficult to remove. The opaque, coil-spring type is normally recommended for use dome-downwards. However the decision as to which way up the cap is fitted is crucial only in a few patients.

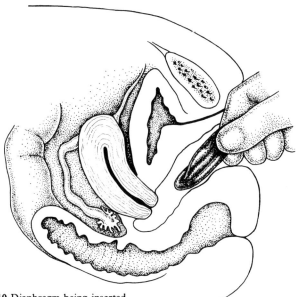

Fig. 8.10 Diaphragm being inserted.

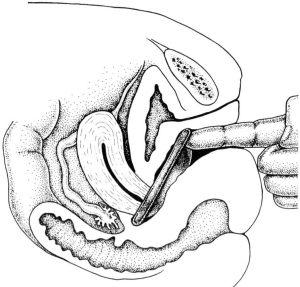

Fig. 8.11 Checking the position of the diaphragm.

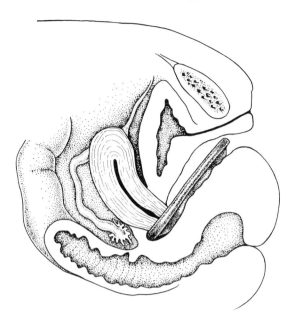

Fig. 8.12 a. Diaphragm too large.

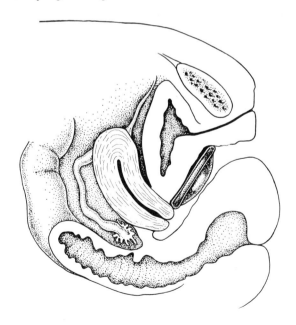

b. Diaphragm too small.

9. The diaphragm is removed by hooking the index finger under the anterior rim and pulling gently downwards.

Teaching the patient. 1. The woman squatting or with one foot on a chair or occasionally lying on her back, according to her preference, is first taught to feel the cervix (Fig. 8.13). Showing her how to do this on a model often helps.

Fig. 8.13 Positions for inserting a diaphragm.

2. The instructor then inserts the cap for the patient, allowing her to feel her cervix covered with the thin rubber. This is very important since correct placing of the cap over the cervix is vital to its success.

3. The patient then removes the diaphragm by hooking her finger under the anterior rim and pulling downwards (Fig. 8.14).

4. She is then taught to insert the cap herself, the instructions being precisely those given for fitting by the doctor or nurse. Emphasis is placed on the downward and backward direction in which the compressed cap is inserted into the vagina. After releasing the cap, the correct covering of the cervix is rechecked (Fig. 8.15), as is the snug fit of the anterior rim behind the pubic arch.

5. Variation in the order of teaching these techniques may be required for some patients depending on their aptitude.

6. If the patient repeatedly inserts the cap into the anterior fornix this can often be overcome by:

    a. the use of a larger cap,

    b. the use of an introducer (Fig. 8.16), or an arcing diaphragm where the difficulty is due to the length of the cervix,

    c. the partner being willing and able to learn the technique of insertion on the patient's behalf,

    d. allowing the woman to teach herself in privacy with the aid of a hand cassette instructor.

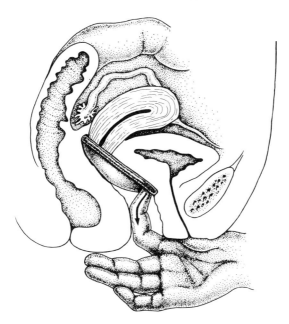

Fig. 8.14 Patient removing a diaphragm. (standing position)

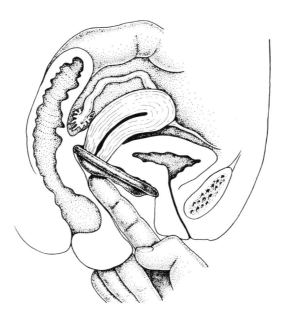

Fig. 8.15 Checking a diaphragm covering the cervix. (standing position)

Fig. 8.16 Introducer.

Cervical cap

Selection and fitting 1. The correct size is that which allows the rim to touch the vaginal fornices easily without a gap; comfortably accommodates the cervix, and is not displaced when the patient bears down.

2. With the patient in the supine position the labia are separated, the rim of the cap is compressed and then guided along the posterior vaginal wall until the posterior rim is just behind the cervix. The thumb and first two fingers are used as illustrated in Figure 8.17.

3. The cap is allowed to open by removing the thumb and then it is pushed upwards onto the cervix with the fingertips (Fig. 8.18). A final check is made to ensure that the cervix is palpable through the bowl and that no gap is left above the rim.

4. The cervical cap is removed by inserting a fingertip between the rim of the cap and the cervix, easing the cap downwards and withdrawing it with the index and middle fingers.

Teaching the patient. The woman is taught to feel her cervix and to insert and remove the cap according to the instructions given above for fitting. In time an experienced user will often develop her own formula for inserting and removing a cervical cap and, provided it is effective, she should be allowed to continue in her own way.

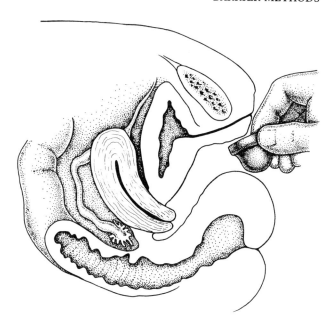

Fig. 8.17 Cervical cap being inserted.

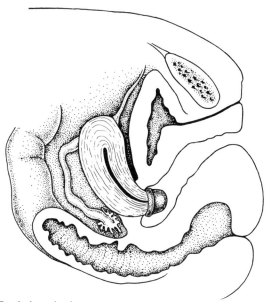

Fig. 8.18 Cervical cap in situ.

Vault cap and vimule

Selection and fitting. 1. The fitting of both these caps is precisely the same as for the cervical cap with modification only of the siting of the upper rim.

2. The correct size for the patient is that which covers the cervix without exerting pressure upon it, while fitting snugly into the vaginal vault (Figs 8.19 and 8.20).

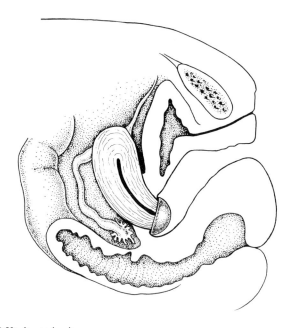

Fig. 8.19 Vault cap in situ.

3. The vimule in particular often answers the problems posed by a difficult fitting.

Teaching the patient. 1. Patients are taught to use these caps in exactly the same way as doctors fit them (see above and p. 170).

2. The string attached to a new vimule is to facilitate removal in the learning period. It is not a permanent feature and should be removed when the user gains confidence.

Schedule of visits

It is customary, but not mandatory, to provide a woman with a practice

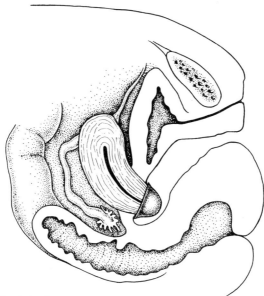

Fig. 8.20 Vimule in situ.

cap for one week. This allows her to gain confidence in the technique of insertion, removal and care of the cap. She can test whether it is comfortable during normal activities including defaecation and micturition.

The patient must understand that the cap should not be used as a contraceptive during the practice week.

A spermicide may also be tried during this week and can be changed at the next visit if any difficulties have arisen. Above all the second visit provides an opportunity to recheck the size and fitting of the cap.

First visit. After the patient is fitted with and taught how to insert the cap its routine use is discussed.

1. Before insertion a spermicidal cream or jelly is applied to the cap. For optimum effectiveness the cap plus a spermicide should be recommended.

2. The cap may be inserted at any convenient time before intercourse to minimise the loss of spontaneity.

3. Additional spermicide should be inserted into the vagina just before intercourse, if the cap has been in place for more than two hours.

4. The cap should be left in position for six hours after the last act of coitus.

5. After removal the cap is washed in warm, soapy water and dried. It may then be powdered with unscented talcum powder. If powder is applied it should be washed off before reinsertion.

6. The cap should be restored to its normal rounded shape and stored in its container in a cool place.

Second visit. 1. The size and comfort of the cap is checked and the routine for its use rediscussed.

2. Exact instructions for applying the spermicide are given:

a. A ribbon of cream or jelly approximately 8 cm long is placed on the upper surface of the diaphragm.

b. One-third of the bowl of the cervical, vault or vimule cap is filled with spermicide. None is used on the rim since this would decrease suction and the cap would tend to slip.

c. Immediately before intercourse a further dose of non-oil based spermicide, of the patient's choice, is recommended for those using caps other than the diaphragm.

Future visits. 1. After three months and then annually.

2. More frequently if difficulties occur or if the woman loses or gains more than 4 kg.

3. After pregnancy or vaginal operation.

Arrangements for visits should not be inflexible — the final frequency of visits will be that which is acceptable to the individual patient.

Instruction in the use of an occlusive cap should inspire confidence. If any patient shows an interest in this method, she should be allowed to experience an initial fitting by a skilled doctor or nurse, just to see how simple and painless the procedure can be.

NEW BARRIERS

New barrier methods are still being developed:

1. A one-sized diaphragm (60 mm), which requires no fitting and may be bought across the counter, is being tried.

2. A cervical cap (Contracap) designed from a plaster cast of the cervix to fit each individual woman is being tested. It can be kept in place for a long time since a one-way valve system opens outwards to allow the escape of secretions and menstrual discharge.

3. A vaginal contraceptive sponge (p. 182).

HOME-MADE BARRIERS

Improvised barriers can be made from tampons, sponges or pads of clean cloth soaked in spermicide, provided that an integral thread ensures their complete removal after use. Such a barrier is inserted deep into the vagina to cover the cervix and should remain in situ for six hours after intercourse.

1. Ideally, manufactured spermicide should be used, but it is often the very absence of forethought in purchasing any contraceptive that forces a couple to use this method.

2. Domestically available spermicides such as vinegar or lemon juice diluted 1/20 with water, mild soap or edible oil are marginally better than nothing.

3. Detergents and strong chemicals are contraindicated since they irritate the vaginal mucosa.

CONDOMS

Since the description in 1544 by Fallopio of the use of a linen sheath as protection against venereal disease, the male protective, barrier or condom has become the most widely used contraceptive in the world. It is recognised by such familiar names as French letter, sheath, rubber, protective, etc. This method should always be included in any discussion on the choice of contraceptive.

Types

1. Most condoms are now made of fine, latex rubber and consist of a circular cylinder (3.0–3.5 cm diameter, 15–20 cm long, 0.03–0.07 mm thick) with one closed, plain or teat-shaped end and an integral rim at the open end. They are normally supplied in one size only, packaged individually, rolled to the rim and hermetically sealed in foil.

2. Lubricated, spermicide-incorporating, coloured and antiallergic variations have been introduced in an effort to improve acceptability.

3. Washable sheaths made of fairly thick rubber are difficult to obtain and are seldom used nowadays.

4. Sheaths made from animal tissue (Fourex) are very expensive.

5. Glans-covering condoms (American tips) are not advocated. They slip off or may be too tight.

6. Urethral insertion of small bags (Gamic devices) is potentially harmful and unreliable.

Mode of action

Like all barrier methods, condoms prevent spermatozoa from reaching the upper female genital tract.

Effectiveness

This is greatest in the experienced user. For couples in the United Kingdom using this method, a use-failure rate of 4 per 100 women years is probably the most accurate figure available.

The commonest reason for failure is improper use, particularly failure to put on the condom at the beginning of intercourse or applying it only just before ejaculation.

Quality control, as with diaphragms, is strict in the United Kingdom and many other countries. Provided condoms are stored in a cool place, used carefully and before their expiry date, failure from leakage through small perforations is highly improbable.

Unpredictable bursting of a condom does sometimes occur. This possibility may be the reason for advocating the use of spermicides at the same time although the increase in effectiveness gained is marginal. Oil-based products (be they lubricant or spermicidal) should be avoided as there is a theoretical risk that they may reduce the tensile strength.

Indications

1. During the period of instruction in the use of a cap.
2. Following delivery, before another method is adopted.
3. Where an additional contraceptive is required while using another method, e.g. during short courses of antibiotics in women on the oral contraceptive pill.
4. Where other methods are unacceptable.
5. Where a couple wish the man to take responsibility for contraception.

Contraindications

1. Where they are psychologically unacceptable.
2. Any anatomical malformation of the penis.
3. Where allergy to rubber develops in either husband or wife.

Advantages

1. Effective when used properly.
2. Easily obtained and fairly cheap.
3. Very few medical complications.
4. Professional help not required.
5. Male method for those who prefer this.
6. Some protection against VD and possibly carcinoma of the cervix.
7. Need for only short-term motivation.
8. Improve the performance in some patients with premature ejaculation.

Disadvantages

1. Interruption of love-making may be unacceptable to some couples, while others can incorporate the use of a condom into love play very satisfactorily.
2. Unaesthetic.
3. Loss of pleasurable sensation.
4. Erectile difficulty may be increased, though some men in later years find the use of the condom helps to maintain an erection.

Availability

Condoms may be purchased from chemists' and barbers' shops, slot machines and other retail outlets, or by mail order. They are not available free on prescription from general practitioners but are free at most family planning clinics. This anomalous situation may prevent their optimal use (Howard & Whittaker 1980).

Instructions for use

1. Unless the condom is lubricated with a spermicide, the woman should insert a spermicidal cream, jelly or pessary into the vagina before intercourse.
2. The condom is unrolled onto the erect penis before any contact with the vulva is made, leaving the tip of the condom empty to accommodate the ejaculate.
3. During withdrawal the condom should be held firmly at the base of the penis so that it remains in place until after the penis has been withdrawn.
4. The penis should be washed thoroughly before any further contact with the woman occurs.

5. Disposable condoms must not be reused.

6. Condoms should not be used after the expiry date marked on the packet.

7. Vaseline or other grease should not be used as a lubricant.

SPERMICIDES

These contraceptives, known internationally as vagitories, comprise a chemical capable of destroying sperm, incorporated into an inert base.

They are available in a variety of forms and, shortly before intercourse, are inserted as far into the vagina as possible.

Most products are suitable for all purposes and may be chosen in accordance with the patient's or doctor's preference. When a combination of two spermicides is advocated it is wise to check their compatibility, according to pH (FPA Approved List 1981).

Creams and jellies

The chemical is incorporated in a stearate soap base in a cream, or in a water-soluble base in a jelly. Both liquefy at body temperature and rapidly disperse throughout the vagina.

Vaginal suppositories

The base consists of gelatin, glycerine or wax. They are foil packed and easy to handle. Since they spread less easily throughout the vagina, weight for weight they are probably less effective than creams and jellies.

Foaming tablets

These are hard white discs which effervesce on contact with moisture, with the release of spermicide and the formation of CO_2 foam.

Aerosol foams

The spermicide is incorporated in an emulsion of oil and water and is stored under gas pressure in a rigid container. It is released, as required, into an applicator, by pressure on a valve on top of the container.

C-Film

This consists of squares of water-soluble plastic impregnated with the potent spermicide nonoxynol 9 (65–70 mg).

Details of the most commonly used spermicides are contained in Table 8.2.

Mode of action

The action of spermicides is two-fold.
1. The base material physically blocks sperm progression.
2. An active chemical kills sperm without damaging other body tissues.

Effectiveness

No spermicide can be considered to be a really effective contraceptive when used alone. It is meaningless to quote failure rates which have been reported as low as 3 per 100 women years and as high as 28 per 100 women years.

Indications

For use in conjunction with diaphragms, condoms, intrauterine devices and coitus interruptus to increase the effectiveness of these methods.

Contraindications

1. Allergy in either partner.
2. Vegetable oil-based products such as Genexol and Rendells pessaries affect the tensile strength of rubber and should therefore not be used with the diaphragm.
3. Aerosol foam should not be used with the diaphragm because if pressure builds up in the vagina the cap could be displaced.

Advantages

1. No systemic effects — no evidence that toxic absorption occurs via the vaginal mucosa, which might prove harmful to the woman.
2. Readily available without prescription.
3. Provide extra lubrication.

Table 8.2 Spermicides

Product	Manufacturer	pH	Chemical constituents
JELLIES			
Duragel	LRC Products	6.0–7.0	Nonoxynol 9, (2%)
Ortho-Gynol Jelly	Ortho-Cilag Pharmaceutical	4.5	Ricinoleic acid, p-di-isobutyl, phenoxypolyethoxye:hanol 1 %
Staycept Jelly	Syntex Pharmaceuticals	4.25–4.75	Octoxynol (polyoxyethylene octyl phenol) 1 %
CREAMS			
Duracreme	LRC Products	6.0–7.0	Nonoxynol 9 (2 %)
Orthocreme	Ortho-Cilag Pharmaceutical	<6.0	Ricinoleic acid, sodium lauryl sulphate, Nonoxynol 9, (2 %)
PESSARIES			
Orthoforms	Ortho-Cilag Pharmaceutical	4.0–5.0	Nonoxynol 9, (5 %)
Staycept Pessaries	Syntex Pharmaceuticals	4.25–5.25	Nonoxynol 9, (6 %)
Genexol	W J Rendell	5.0–6.0	p-tri-isopropyl phenoxy decaethoxy ethanol (Texafor FN11) 0.08 g. Fractionated palm kernel oil B.P., 1.52 g
Rendell	W J Rendell	6.0	Nonylphenol ethylene oxide, 0.08 g. Fractionated palm kernel oil B.P., 1.52 g
FOAMS			
Delfen Foam	Ortho-Cilag Pharmaceutical	4.5–5.0	Nonoxynol 9, (12.5 %)
Emko Foam	Syntex Pharmaceuticals	7.4–7.8	Benzethonium chloride, (0.2 %,) Nonoxynol 9, (8 %)
C-Film	Potter and Clarke	5–7	Nonoxynol 9, 65–70 mg

Disadvantages

1. Relatively high failure rate.
2. Aesthetically unacceptable and messy.
3. Have to be inserted shortly before intercourse.
4. Vaginal suppositories are unsuitable for use in tropical temperatures.
5. Aerosol preparations may be difficult to use if the container is not shaken properly.
6. Occasional complaints of an unpleasant odour, stinging or discomfort in the vagina.

The suggestion that there was an increased risk of fetal abnormality and early abortion if the woman became pregnant while using spermicides as a contraceptive has not been confirmed (Huggins et al 1982).

Availability

Spermicides may be purchased from chemists and other shops and by mail order. They are available free on prescription from general practitioners and at family planning clinics.

Instructions to patients

1. Cream and jellies may be inserted into the vagina with an applicator 2–3 minutes before intercourse, or on an occlusive pessary (p. 174). A dose of 2 g is adequate.
2. Foaming tablets are slightly moistened and inserted high into the vagina 3–10 minutes before intercourse.
3. Vaginal suppositories are inserted about 15 minutes before intercourse.
4. One applicator full of foam should be inserted into the vagina just before intercourse (Fig. 8.21). The aerosol can is shaken well to ensure adequate mixture of the spermicide and the foam. Thereafter the applicator should be washed in warm water and thoroughly dried.
5. C-Film is inserted, with the fingers, high into the vagina at least 30 minutes before intercourse. Alternatively it may be placed over the glans penis and transferred into the vagina in this way.
6. If spermicide is inserted more than 2 hours before intercourse takes place, a second dose of spermicide should be used.

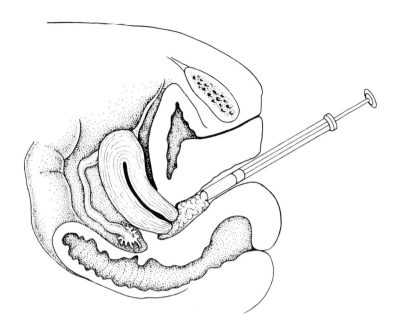

Fig. 8.21 Foam insertion.

THE VAGINAL CONTRACEPTIVE SPONGE

This is a soft white circular sponge 5.5 cm in diameter made of polyurethane foam and impregnated with the spermicide nonoxynol-9. The woman inserts it high into the vagina and a polyester loop is attached to facilitate removal (Fig. 8.22).

Marketed under the name 'Today', it has been approved by the Committee on Safety of Medicines and will soon be available without prescription. A box of three sponges will cost approx. £2.00.

Mode of action

It acts as a barrier to the passage of sperm into the cervical canal and the nonoxynol-9 destroys the sperm.

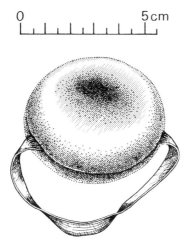

Fig. 8.22 Vaginal contraceptive sponge.

Effectiveness

Although a trial at the Margaret Pyke Centre in London found the sponge to be much less reliable than the diaphragm, studies in five other countries have found it to be almost as effective as other vaginal methods.

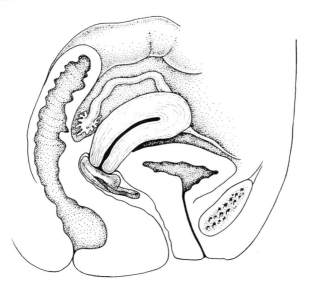

Fig. 8.23 Sponge in situ.

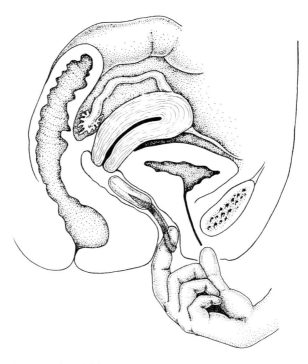

Fig. 8.24 Sponge being withdrawn.

Indications

For the woman who wants to take the responsibility for her own contraception but wishes neither to use 'medical methods' nor seek medical help with fitting a barrier.

Contraindications

Allergy to nonoxynol-9 is the only contraindication recognised so far. Others may come to light when the sponge is more widely used.

Advantages

1. One size is suitable for most women.
2. No fitting is required.
3. It is less messy and more comfortable than other barriers.
4. There is no need to add additional spermicide before each act of intercourse.
5. There is no evidence of any significant health risk.

Disadvantages

1. The low failure rate claimed in the pilot studies may not be substantiated when the sponge becomes freely available for general use.

2. A very small number of women or their partners may be sensitive to the spermicide.

3. Concern by some doctors about any such foreign body left in the vagina for 24 hours or more, particularly the risk of toxic shock syndrome.

Instructions for use

1. The sponge is thoroughly moistened and inserted high into the vagina. (Fig. 8.23)

2. It may be worn for up to a total of 24 hours. (In America a licence has been granted for 48 hours use.)

3. Intercourse may be repeated as often as desired during this time.

4. The sponge should be kept in place for at least 6 hours after the last intercourse, removed and thrown away (Fig. 8.24)

5. Some doctors recommend that it should not be used during menstruation since clinical tests of the product have not been extensive enough to assess the risk of developing toxic shock syndrome. Yet the manufacturer's literature contains no such warnings. It says that the sponge *may* be used during menstruation but that it should be inserted immediately prior to love-making and always removed six hours afterwards.

6. It may be left in place while swimming and having a bath or shower.

COITUS INTERRUPTUS

This is the oldest method of birth control, widely used in Christian and Muslim communities but less so in oriental countries. In Jewish writings 'the spilling of the seed' was considered sinful and the sin of Onan (*Genesis* 38, verse 9) probably referred to coitus interruptus. On this St Augustine based his condemnation of contraception leading eventually to the doctrine set out in *Humanae Vitae*.

Doctors should ensure that realistic advice on this method is available, free from the considerable adverse comment usually made about it in family planning literature. This applies particularly to general practitioners whose opportunity to discuss birth control is generally less rehearsed than in a family planning clinic.

Description

Coitus interruptus is withdrawal of the erect penis from the vagina before ejaculation. It is described by users as 'being careful', 'withdrawal' and by local euphemisms implying stopping before the effective 'end of the line'. Such terms should be recognised to avoid a false history of 'no previous contraception'.

The method requires discipline on the part of the male and is practised most successfully by those who recognise the imminence of climax and can withdraw quickly before ejaculating. Many loving partnerships take pride in the control exercised by the male partner and then happily complete sexual satisfaction for the female by continuing manual stimulation to climax.

Mode of action

Since ejaculation occurs outside the vagina, semen is not deposited within the vagina and pregnancy should not ensue. However, a drop of seminal fluid containing thousands of sperms may escape from the urethra before ejaculation and may result in conception.

Effectiveness

Failure rates vary with the age and experience of the couple and range from 5 to 20 per 100 women years.

Precise figures are difficult to obtain but the few available studies endorse the historical evidence that this is still a widely used method. In the United Kingdom in 1978, 700 000 couples were known to be using this method compared to 600 000 couples using the IUD and 300 000 the diaphragm (Ch. 13).

Family planning clinics are naturally influenced by the fact that successful and contented users seldom consult them, whereas those who fail, complain and seek alternative advice.

Indications

In an emergency situation where intercourse takes place and no other contraceptive is available.

Contraindications

Where a reliable and effective means of preventing pregnancy is desired.

Advantages

1. No equipment or preparation required.
2. No medical supervision necessary.
3. Costs nothing.
4. No evidence of either psychological or physical side effects.
5. Allows total privacy about the couple's sexual relationship.

Disadvantages

1. High failure rate when practised by sexually inexperienced men, especially the young.
2. Limits enjoyment of sexual intercourse.

COITUS RESERVATUS

This involves vaginal penetration, but restriction of movement thereafter so that intravaginal stimulation does not lead to ejaculation, either within or outside the vagina.

This technique is probably more difficult and less reliable, save for those couples whose love-making techniques are somewhat passive.

COITUS INTERFEMORA

The stimulation of the erect penis and ejaculation between the thighs of the woman instead of in the vagina may be an easier and sometimes acceptable alternative.

COITUS DESIGNED TO AVOID THE VAGINA

Oral and anal sex are increasingly practised. While neither is advocated, contraceptive counsellors should expect the possibility of their discussion, especially in relation to sexually transmitted diseases.

RISKS/BENEFITS

Choice of barrier methods depends on acceptance of the fact that the advantages of simplicity and lack of side effects outweigh the desire for maximum efficiency. They depend on consistent and practised use for maximum effectiveness. Already established confident users, especially of occlusive caps and sheaths, need not be encouraged to change to more reliable modern methods unless there is some increased need or different pattern of sexual activity.

Nonetheless, in advising the young, fertile and highly sexually active group to whom accidental pregnancy would be a disaster, doctors should be cautious about encouraging their use.

Failures with occlusive methods (cap and condom) do occur. Patients may be at risk of pregnancy following premature removal or rupture of a sheath or dislodgement of a cap. In these situations, postcoital contraception can be used and couples using barrier methods should be aware of its availability (see Ch. 11).

REFERENCES

Bone M R 1980 Health trends, DHSS, 12. 4: 87
Family Planning Association 1981 Approved list of spermicides. FPA, London
Howard G, Whittaker I 1981 Difficulties in obtaining condoms on the NHS.
  British Journal of Family Planning 7: 12–15
Huggins G, Vessey M, Flavel R et al 1982 Vaginal spermicides and outcome of
  pregnancy; findings in a large cohort study. Contraception 25 (3): 219–230
Kelaghan J, Rubin G L, Ory H W, Layde P M 1982 Barrier-method contraceptives
  and pelvic inflammatory disease. Journal of the American Medical Association
  248 (2): 184–187
Vessey M, Wiggins P 1974 Use-effectiveness of the diaphragm in a selected family
  planning clinic population in the UK. Contraception 9: 15–21
Wright N H, Vessey M P, Kenward B, MacPherson K, Doll R 1978 Neoplasia
  and dysplasia of cervix uteri and contraception: a possible protective effect of the
  diaphragm. British Journal of Cancer 38: 273–279

Natural family planning

The traditional terms 'rhythm' and 'safe period' to designate methods of birth control based on periodic abstinence have been replaced over the past two decades by the term 'natural methods of family planning' (NFP). These methods are based on the recognition of the naturally occurring signs and symptoms of ovulation and the fertile and infertile phases of the menstrual cycle.

Abstinence from intercourse during part of the menstrual cycle to avoid conception has been practised throughout the ages but has only been established on a sound scientific basis over the past 50 years. Ogino and Knaus in 1930 showed that conception can only occur during a short period around the time of ovulation and that, regardless of the length of a woman's cycle, the interval between ovulation and the next menstruation is fairly constant — approximately 14 days. This information forms the basis of the calendar or rhythm method.

Recognition that the shift in basal body temperature is related to the activity of the corpus luteum after ovulation forms the basis of the thermal or temperature method first described by Férin.

In 1964 Drs John and Evelyn Billings innovated the 'ovulation method' in which the changes in cervical mucus at different times in the cycle are used to indicate fertile and infertile phases.

Although periodic abstinence was condemned by the early Roman Catholic church, the Papal encyclicals *Casti Conubii* (1930) and *Humanae Vitae* (1968) both approved it.

SCIENTIFIC BASIS

Ovulation

Ovulation, the essential event which predetermines the beginning and the end of the fertile period can be detected only by indirect methods. The main ones being used are:

1. The concentration of the ovarian hormones, oestrogen and progesterone and those from the pituitary, follicle stimulating hormone (FSH) and luteinising hormone (LH) in the peripheral blood.

2. The concentration of metabolites of these hormones in the urine.

3. The effects of oestrogen and progesterone on target organs in the body (the cervix and vagina) and on basal body temperature.

The first and second methods are not currently used in natural family planning but form the basis of research into new techniques for the future.

Recent developments using the technique of ultrasound scanning, have shown that by serial ultrasonic scans, one can monitor the growth of the Graafian follicle from a diameter of 5 mm to an average maximum diameter of 23 mm and subsequently detect the development of a corpus luteum (Kerin et al 1981). It is probable that this method, still in its infancy and as yet restricted to centres of ultrasound expertise, will become in the future the focal point against which all the indirect hormonal and clinical indicators in use today will be evaluated.

Figure 9.1 relates to hormone levels during the menstrual cycle, to the development of the endometrium and changes in the cervix and basal body temperature. The growing ovarian follicle, stimulated by FSH, produces increasing amounts of oestrogen which cause the cervix to shift upward in the pelvic cavity to become softer and the os to open. Oestrogen also causes the production of increased amounts of slippery, stretchy mucus which reaches peak proportions at about the same time as the peak of oestrogen in blood and urine just prior to ovulation. After ovulation the increased amounts of progesterone produced by the corpus luteum reverse the effect of oestrogen on the cervix and also raise the basal body temperature.

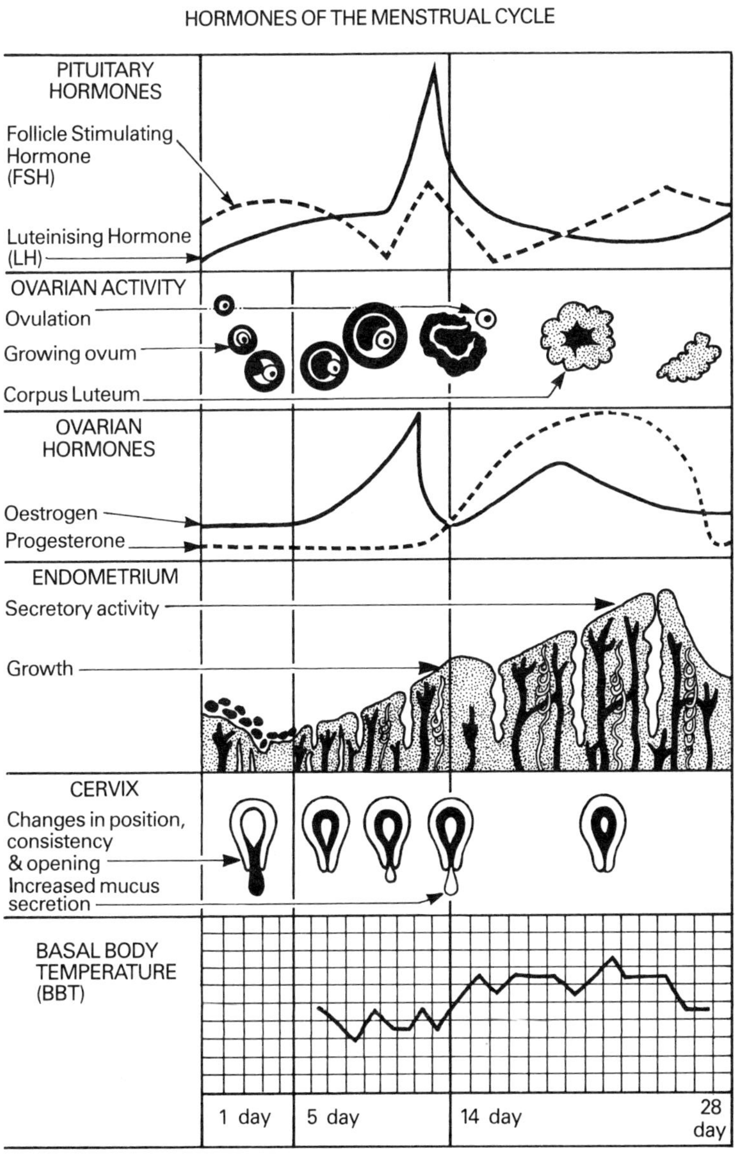

HORMONES OF THE MENSTRUAL CYCLE

Fig. 9.1 Hormones of the menstrual cycle.

The sperm and the ovum

The ovum remains capable of being fertilised for only 24 hours after release from the ovary. It is much more difficult to be precise about sperm survival since motility is a poor indicator of the ability to fertilise an ovum and much depends on the environment. The average time for survival of sperm is probably three days but there is evidence that in some instances they may survive for up to seven days.

METHODS

Calendar method

This relies on:

1. Ovulation occurring 12–16 days before the beginning of the next period.

2. The probability of working out the fertile days in the cycle on the basis of information recorded in previous cycles.

3. Making allowances for the survival time of ovum and sperm.

Instructions

Several formulae can be used to calculate the fertile days in the cycle. One is as follows:

1. Record cycle length for six months, e.g. 25 28 26 30 26 28 days.

2. Subtract 20 days from the shortest cycle to calculate the first fertile day.

3. Subtract 10 days from the longest cycle to calculate the last fertile day.

4. The fertile phase is then from day 5 (25–20) to day 20 (30–10)

Abstinence should therefore be observed from day 5 to day 20 inclusive.

This calculation is never used today in NFP Clinics. Commercially available calculators are no more reliable than the above calculation.

Basal body temperature method

This method relies on the fact that progesterone from the corpus luteum raises the temperature of the body at complete rest by 0.2–0.4°C and maintains it at this level until the onset of the next menstruation.

Instructions

Taking the temperature. 1. If possible use a special ovulation thermometer graduated from 36–38°C as it is easier to read.

2. Record the temperature at the same time each day immediately after waking, before getting out of bed or having anything to eat or drink.

3. Keep the thermometer in the mouth for five minutes. Rectal and vaginal recordings each for three minutes are sometimes recommended but many women find them distasteful.

4. Record the temperatures on the chart in the centre of the appropriate square.

5. Record any intercurrent illness e.g. colds or flu and any medication taken.

6. Begin a new chart on the first day of each period.

Interpretation

Ten per cent of charts are difficult to interpret but users get better at this with experience.

1. The infertile phase begins as soon as three consecutive daily temperatures above the level of the previous six consecutive daily temperatures (excluding days 1–4 of the cycle) have been recorded.

Occasionally a spike temperature of no more than 0.2° above that recorded on the days on either side is registered. This may be due to a cold, to having taken alcohol the night before or to recording the temperature at a different time of the day. One such spike may be ignored.

2. The couple can have unprotected intercourse from the 3rd day of temperature rise until the beginning of the next menstrual period.

3. In the next cycle they must not have unprotected intercourse between the first day of the period and the third day on which the temperature is recorded at the higher level.

Ovulation method

Also called cervical mucus and Billings method (Billings & Billings 1973). This method relies on recognising the changes that occur in cervical mucus under the influence of oestrogen and progesterone at different times in the menstrual cycle (p. 191). Four phases can be recognised:

1. The dry days following menstruation when the mucus forms a thick viscid plug in the cervix and the vulva and vaginal entrance feel dry.

2. Days in the early preovulatory phase when mucus first appears. The dry sensation disappears. The mucus is cloudy, white and of sticky consistency.

3. The wet days immediately prior to ovulation when mucus becomes more copious, clear, slippery and with the consistency of egg white. The vulva feels moist or even wet. If midcycle bleeding occurs the mucus may have a pink tinge. These days mark the peak of fertility.

4. The days following the peak during which mucus becomes sticky, cloudy and scanty and the vaginal entrance and vulva again feel dry. This phase persists until the next menstrual period.

Instructions

1. Observations should be made at a convenient time each day, such as when going to the toilet, and recorded in the evening on the appropriate chart either by using identification stickers or letters such as P for periods, D for dry, M for moist (Fig. 9.2).

2. The state of the mucus is noted at micturition. The woman does this by wiping the vulva with toilet paper and noting the characteristics of the mucus present. She should also record subjective feelings of wetness and dryness. (Fig. 9.2)

3. The peak mucus day is marked with a cross. It is the last day of the slippery, elastic watery mucus or a sensation of wetness. It can only be identified retrospectively.

4. The following days are numbered day 1, day 2, day 3 and day 4.

Interpretation

1. Intercourse can take place on alternate dry days following menstruation.

2. As soon as the first sign of mucus appears the couple must abstain.

3. Sexual intercourse should be avoided until the evening of the 4th day after the peak mucus.

Symptothermal Methods

These methods seek to combine several clinical indices of ovarian function in order to pinpoint the fertile period with greater precision

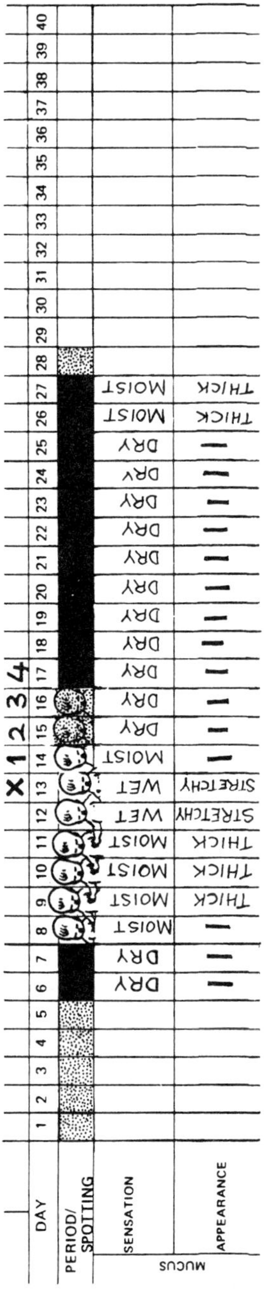

DAY	1	2	3	4	5	6	7	8	9	10	11	12	13	14	15	16	17	18	19	20	21	22	23	24	25	26	27	28	29	30	31	32	33	34	35	36	37	38	39	40
PERIOD/SPOTTING																																								
MUCUS SENSATION						DRY	DRY	MOIST	MOIST	MOIST	MOIST	WET	WET	MOIST	DRY	DRY	DRY	DRY	DRY	DRY	DRY	DRY	DRY	DRY	DRY	MOIST	MOIST													
MUCUS APPEARANCE						—	—	—	THICK	THICK	THICK	STRETCHY	STRETCHY	—	—	—	—	—	—	—	—	—	—	—	—	THICK	THICK													

Fig. 9.2 Ovulation method.

and reliability. This approach combines on a single chart calculation and temperature methods with observations of the cervical mucus, the cervix itself (dilatation, softness and position), midcycle pain and bleeding and breast sensitivity (Fig. 9.3). Special emphasis is placed on the predominant symptom for any particular woman, thus tailoring the method to the individual.

EFFECTIVENESS

1. The calendar method is unreliable and should not be recommended for use on its own.

2. The temperature method, if properly used, can be effective but, since intercourse is restricted to the postovulatory phase, it is seldom used alone but may be combined with other techniques.

3. Claims that the ovulation method is effective in 98% of couples have to be put in perspective. Such effectiveness is only attained by those who are very highly motivated. In the WHO prospective multicentre study (1981) failure rates of 20–25% were recorded in those who did not adhere strictly to the rules.

4. In practice, the symptothermal method has better use-effectiveness than the cervical mucus method alone.

Table 9.1 lists the most important prospective NFP studies using the Pearl Index (failure rate per 100 women years) as a measurement of effectiveness. Use-effectiveness is lower and shows a much wider range than that quoted for other methods of contraception. Three important factors contribute to success or failure.

a. *Motivation:* this is important for all methods of family planning but particularly for coitus-related techniques such as barriers and NFP. Success or failure depends on the joint commitment of the couple to cope with abstinence during the fertile time. Couples using NFP to limit their family do so with much greater success than those who wish only to space or delay the next pregnancy.

b. *Efficient teaching:* success is directly proportional to the users' ability to identify the fertile time during the cycle. It is therefore important that they should be taught well. The couple must understand the physiological principles on which the method is based. This can be achieved, even for people of a low educational level, by using simple terms and analogies.

Success is greater where both partners are taught about fertility awareness. They are then better able to share the responsibility for family planning and accept that modification of their pattern of sexual behaviour is required. When they understand why periods of

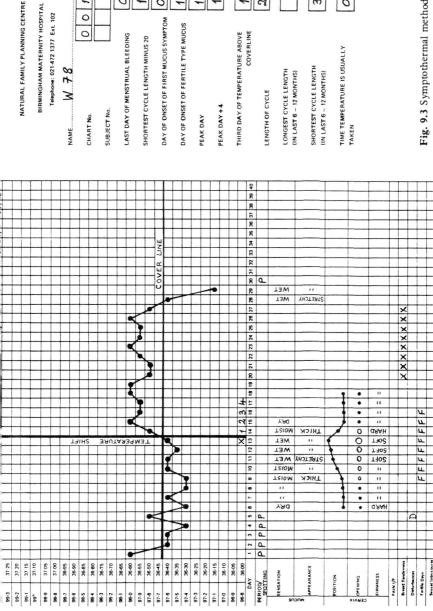

Fig. 9.3 Symptothermal method.

Table 9.1 Prospective studies

Author	Date	Method & usage	No. of participants	Months of use	Pregnancies per 100 women year method failure	Pregnancies per 100 women year use failure
Marshall	1968	Basal body temperatures only postovulatory	321	4739	1.2	6.6
Ball	1976	Billings ovulation pre- & postovulatory	122	1628	2.9	15.5
Parenteau-Carreau, Lanctot & Rice	1976	Symptothermal pre- & postovulatory	101 (limiters)	2278	—	1.1
Parenteau-Carreau, Lanctot & Rice	1976	Symptothermal pre- & postovulatory	67 (spacers)	1286	—	14.9
Johnston, Roberts & Spencer	1978	Symptothermal with calculations pre- & postovulatory	192	3198	—	8.6
Johnston, Roberts & Spencer	1978	Billings ovulation method pre- & postovulatory	586	6525	—	27.6
Rice, Lanctot, Garcia Devesa	1979	Symptothermal pre- & postovulatory	723	14 416	0.75	5.5
Klaus, Egizio,	1979	Billings ovulation	1139	27 336	1.0	20.00
Medina et al	1980	Symptothermal pre- & postovulatory	111	969	—	19.8
		Billings ovulation method pre- & post ovulatory	130	1064	—	33.8
Wade et al	1980	Symptothermal pre- & postovulatory	239	1668	0	13.7
		Billings ovulation method pre- & postovulatory	191	1269	5.7	39.7
WHO Multicentre Prospective Trial	1981	Billings ovulation method pre- & postovulatory	725	7514	2.8	19.9

abstinence are necessary it becomes easier to accept them. Many couples find that, thereafter, the pleasures of intercourse are enhanced.

Teachers themselves must be highly motivated and able to communicate. Each develops his or her own teaching technique depending on the method and the individual couple.

The World Health Organisation and the British Life Assurance Trust for Health Education (1979) have produced a NFP learning package for use by lay teachers rather than by doctors and nurses.

c. *Characteristics of the woman:* some women find it very difficult to identify the particular phase of the cycle from mucus observation. Yet in one study 97% of women were able to identify correctly mucus symptoms in the third cycle following instruction (Burger 1981).

Some women dislike even touching the vulva and the mucus method is inappropriate for them. Others find it difficult to observe periods of abstinence during the fertile phase.

ADVANTAGES

1. No known physical side effects.

2. Normal physiology is not interfered with.

3. Users develop an awareness of fertility which helps promote a more responsible attitude to family planning and, for some, a better marital relationship.

4. Morally and culturally acceptable in any society where a period of abstinence is accepted and other methods of contraception are not.

5. The couple have personal control over their own fertility to avoid or achieve a pregnancy as they wish.

6. Once established as an efficient user (after proper teaching), no further follow up or expense is necessary.

7. Natural methods can be an advantage in helping subfertile couples to achieve a pregnancy.

DISADVANTAGES

1. Although method effectiveness may be high, use-effectiveness is not (p. 198).

2. There has to be a modification or adjustment of sexual behaviour and this requires the co-operation of both partners.

3. The woman has to learn to observe, chart and interpret some signs and symptoms and this takes time.

4. High motivation is required to continue to do this and some women find it such a daily burden that they abandon the method.

5. It is not always easy to find competent teachers.

6. The potential hazard of chromosomal abnormality resulting from the fertilisation of an aged ovum by an aged sperm leading to fetal abnormality or abortion. Recent studies have shown that there is no increased risk (Bonnar 1984).

INDICATIONS

1. The principal indication is for those who for moral, cultural or medical reasons cannot, or do not wish to, use other methods.

2. For those who use barrier methods but wish to restrict their use to the fertile time of the cycle. The fertility awareness component of NFP can help them achieve this.

CONTRAINDICATIONS

1. Where the male partner is unwilling to practise periodic abstinence. It is therefore obvious that NFP is not suitable for those who have casual sex.

2. Women with normal cycles who cannot identify fertility indices (3%).

3. Groups such as adolescents, perimenopausal women and women who have had cervical surgery (cautery or conisation) may experience difficulty. However if they are highly motivated and well taught they can often succeed.

NATURAL FAMILY PLANNING CLINICS

Although doctors and nurses in family planning clinics and general practice are often happy to teach their patients how to use natural methods of family planning, others prefer to send them to recognised teachers who are generally highly motivated lay women many of whom practise the technique themselves and are confident in its use. They pass on their knowledge, enthusiasm, expertise and trust in the method to those whom they seek to help.

During the period of instruction a couple learns to detect the fertile days in the cycle and to adjust their sexual behaviour accordingly.

Teaching is spread over a period of 3 months divided into 6 sessions of approximately 60–90 minutes each. Teaching sessions are conducted either at the clinic, the home of the instructor or of the learner which ever is the most convenient. This approach allows for differential learning rates by couples and affords the teacher an opportunity to supervise charts, etc. over three cycles.

IPPF POLICY

The IPPF policy on NFP (for which the term 'periodic abstinence' is employed) was approved by its Central Council in November 1982. It concludes:

> 'Although theoretical calculations may suggest that these techniques can be relatively effective, in practice the failure rates are high. In recent major studies, almost 20% of women using the symptothermal method became pregnant within a year, as did about 25% of those using the cervical mucus method, compared with less than 5% of those using oral contraceptives or intrauterine devices. While the symptothermal method appears to be more effective than the cervical mucus method, the two methods show wide and overlapping ranges of pregnancy and discontinuation rates among different groups of women. Failure and discontinuation rates at this level are unusually high as compared with other methods. IPPF therefore does not advise that periodic abstinence be considered as an equal alternative to other more effective family planning methods.
>
> Nevertheless, periodic abstinence can be the only choice for individuals and couples who cannot, or do not want to, use other methods of fertility regulation for a variety of reasons. Family planning associations should therefore familiarise themselves with the techniques and be prepared to teach them if a demand can be demonstrated. Couples electing to use periodic abstinence should, however, be clearly informed that the method is not considered an effective method of family planning although individual highly motivated couples may find it effective.
>
> It should be recognised that periodic abstinence as a method is better than no method at all. There are also various benefits to be obtained from an understanding of the reproductive cycle. It provides an opportunity for women to learn about their physiology. The identification of the fertile phase is the basis for one contraceptive practice whereby couples choose to use barrier methods only during those days estimated to be the fertile phase of the woman's cycle. It may also be a starting point for the use of more effective contraception. Finally, methods for the detection of ovulation have been used and continue to be valuable in the diagnosis and, more importantly, in the treatment of infertility.'

Interest in NFP has grown in the past decade and is being reinforced constantly by present day ecological trends and a general concern about the complications of drugs and devices including those used in the field of contraception. However, all natural methods in their present form are still very crude indicators of the fertile time and are associated with many disadvantages. On the credit side, self-awareness of one's fertility and sexuality acquired in learning natural methods can be an important factor in personal development and marital harmony. For some people NFP can provide an equally acceptable alternative to other more effective methods of contraception.

REFERENCES

Billings E L, Billings J J 1973 The idea of the ovulation method. Australian Family Physician 2: 81–85

Bonnar J 1984 Biological methods of identifying the fertile period. The Proceedings of the XIth World, Congress on Fertility and Sterility. Dublin 1983 p 77–92. Editors: R F Harrison, J Bonnar, W Thompson Publishers: M T P Press Limited.

Burger H G 1981 The ovulation method. Proceedings International Seminar in Natural Family Planning, Dublin, October 1979, p 79–90

Clarke W D 1981 The WHO family fertility education learning package. Proceedings International Seminar on Natural Family Planning, Dublin, October 1979, p 110–114

Kerin J F, Edmonds D K, Warnes G M et al 1981 Morphological and functional relations of graafian follicle growth to ovulation in women using ultrasonic, laparoscopic and biochemical measurements. British Journal of Obstetrics and Gynaecology 88: 81–90

Kleinman R L (ed) 1983 Periodic abstinence for family planning. IPPF, London

World Health Organisation 1981 A prospective multicentre trial of the ovulation method of natural family planning II. The effectiveness phase. Fertility and Sterility 36: 591–598

Sterilisation

Surgical methods of fertility control have a long and often bizarre history going back many centuries. In the early 1900s sterilisation, as we now know it, was first performed generally for eugenic reasons such as severe mental retardation.

In recent years widespread acceptance of sterilisation on the basis of personal wish to prevent conception has come about largely due to changes in attitudes. The use of simpler and safer surgical techniques and advances in anaesthesia have also contributed to the increase in its popularity.

In 1972 vasectomy became available under the National Health Service but until recent years it has lagged behind female sterilisation. This was partly because both the public and the medical profession were slower to accept vasectomy and also because of the uncertain and complex legal position.

MALE STERILISATION

Vasectomy is a simple procedure which can be performed under local or general anaesthesia on an out-patient basis. It is quicker and in many ways easier than female sterilisation. The operation is well within the capability of doctors with basic surgical skills and is associated with fewer complications than female sterilisation. It involves the division of the vas deferens on each side, followed by ligation of the divided ends (Fig. 10.1). This may be combined with other procedures, e.g. excision of a small segment of the vas, cauterisation of each end or with separation of the cut ends into separate fascial planes (pp 214 and 215).

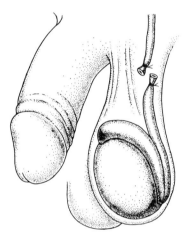

Fig. 10.1 Vasectomy

FEMALE STERILISATION

Sterilisation for purposes of fertility control is usually achieved by blockage of the fallopian tubes carried out by laparoscopy or laparotomy.

Hysterectomy by the abdominal or vaginal route should be reserved (and indeed may be indicated) for those who have menorrhagia or pelvic disease and require sterilisation in addition. The operation carries a much greater risk than simple modern methods of sterilisation.

Laparoscopy

Laparoscopy has now become the technique of choice because of its simplicity, speed, minimal inconvenience, and acceptability to the patient. Although the technique requires skill, once mastered it carries a very low complication rate.

When the laparoscope has been introduced into the peritoneal cavity tubal occlusion may be achieved by the following methods.

Diathermy coagulation

Two small segments of each tube are occluded by diathermy (Fig. 10.2). Caution has to be exercised to prevent damage to the surrounding tissues, a risk which has been reduced by replacing unipolar with bipolar diathermy.

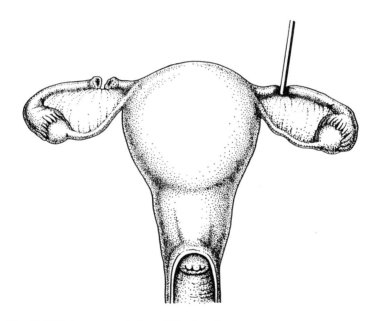

Fig. 10.2 Diathermy coagulation of fallopian tubes.

Application of plastic clips

A specially designed clip is placed over a portion of each tube (Figs 10.3 and 10.4). Some surgeons prefer to place two clips on each side.

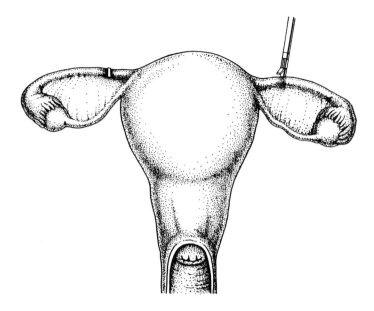

Fig. 10.3 Application of clips

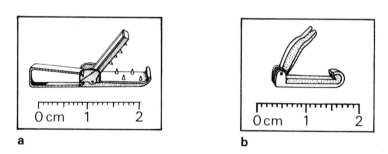

Fig. 10.4 a Hulka clip,   **b** Filshie clip.

Minimal tubal damage is caused and the risk of trauma to surrounding tissues is extremely small.

Application of Falope (Yoon) rings

A small silastic ring is placed over a loop of each tube (Fig. 10.5).

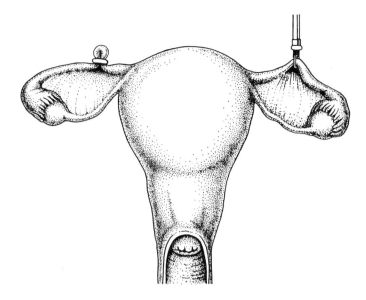

Fig. 10.5 Application of Falope rings.

Laparotomy

Laparotomy is used to carry out tubal ligation, salpingectomy or hysterectomy.

Tubal ligation

1. The Pomeroy operation consists of ligating the 'neck' of a loop of tube with absorbable suture material followed by excision of the loop (Fig. 10.6). When the sutures have been absorbed the cut ends tend to separate giving a low failure rate.

2. Modification of the Pomeroy operation includes separating the two cut ends by interposing the round ligament between them, burying one end under the peritoneum of the broad ligament or overlapping the cut ends.

3. The Madlener technique consists of crushing and ligating a loop of tube with nonabsorbable suture material (Fig. 10.7).

4. The Irving technique consists of excision of a portion of tube and burying the medial end back on itself into the uterine cornu (Fig. 10.8).

These operations may sometimes be performed through a small midline incision (minilaparotomy).

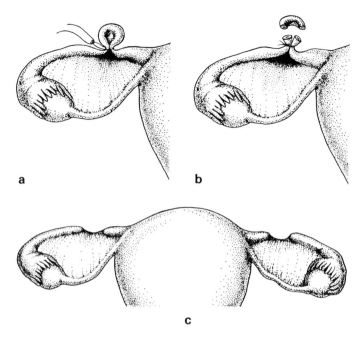

Fig. 10.6 Pomeroy technique.

Other techniques

Other techniques which are not recommended include:

1. Total salpingectomy — this is highly effective but carries a greater risk of complications, particularly haemorrhage.

2. Culdoscopy — this involves endoscopic sterilisation via the vaginal route and has been supplanted by laparoscopy.

3. Hysteroscopy — through the hysteroscope the tubes are occluded at the cornual ends using coagulation or chemicals.

4. Irradiation of the ovaries — this provides effective sterilisation but is very rarely used nowadays, being reserved for those few patients whose medical condition makes any operative procedure dangerous.

EFFECTIVENESS

It is difficult to give accurate figures for the failure rate of various methods of sterilisation. Much depends on the technique used, the expertise of the surgeon and for how long the cases have been followed up afterwards.

Vasectomy is generally accepted as being more effective than female

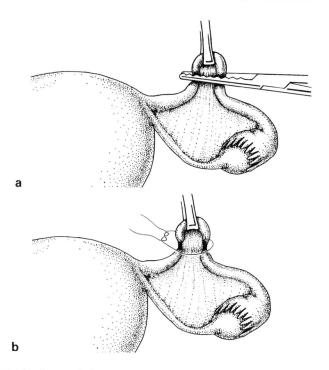

a

b

Fig. 10.7 Madlener technique.

sterilisation. A failure rate of 0.02 per 100 women years for vasectomy and 0.13 per 100 women years for female sterilisation is reported in the Oxford/FPA study (Vessey et al 1982), but other studies quote higher failure rates.

Failure rates of 0.1–3% have been reported in women sterilised by laparoscopic techniques. The failure rate with Hulka clips tends to be higher than with Falope rings or cautery.

In the Oxford/FPA study, after 7 years follow-up, 1% of all the women who had been sterilised (by a variety of techniques) had become pregnant.

Pomeroy and Irving techniques have a very low failure rate but require laparotomy.

INDICATIONS

1. For the couple who are certain, for whatever reason, that their family is complete and who are as sure as they possibly can be that they do not want another pregnancy.

2. For the individual or couple who are carriers of an inherited disorder which they do not wish to risk transmitting to their offspring.

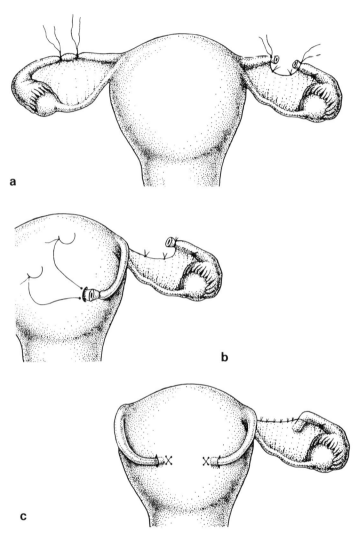

Fig. 10.8 Irving technique

CONTRAINDICATIONS

1. *Marital problems.* Sterilisation should not be considered as an attempt to solve marital disharmony or psychosexual problems unless

it can be shown that these are a direct result of the fear of further pregnancy. In practice this is very rarely the case, as these problems are usually complex and multifactorial. If the marriage breaks up it is possible that a new relationship may be entered into with subsequent desire for another baby.

2. *Uncertainty about the finality of the procedure.* Sterilisation should not be carried out in couples who believe that the operation may be readily reversed at some future date, should they change their minds and want another baby.

3. *Psychiatric illness in either partner.* A psychiatric opinion should be sought before a decision is reached.

4. *Serious physical disability.* The risks of the procedure itself may constitute a contraindication and it may be more appropriate for the other partner to be sterilised. The life expectancy of the ill person must also be taken into account.

5. *Young couples.* Requests for sterilisation by couples in their teens and early 20s must be carefully assessed since regret and requests for reversal of the procedure are commoner among this age group.

ADVANTAGES

Both male and female sterilisation provide highly effective long-term contraception without the need for continued motivation or positive action. Side-effects from continued use of some reversible methods are avoided.

The advantages of male as opposed to female sterilisation are:

1. It is easily performed by a surgically trained practitioner.

2. It can be performed under local analgesia as an outpatient in a surgery or clinic.

3. It requires less sophisticated equipment and is cheaper.

4. It usually involves less disruption to the man's everyday life than does female sterilisation to the woman's.

5. Mortality and significant operative morbidity are virtually nonexistent.

The advantages of female as opposed to male sterilisation are:

1. It is immediately effective, unlike vasectomy which may take several months to be absolute.

2. In many couples it is the female who is more motivated towards using permanent contraception particularly if she has had to suffer the unwanted effects of alternative methods.

DISADVANTAGES

1. Female sterilisation carries a risk of operative mortality and morbidity (p. 220)

2. Should the patient change his or her mind and desire another child the operation can be reversed even by experienced surgeons in less than 60% of cases.

3. Sterilisation is more complicated than alternative contraception and requires specialised skills and facilities.

4. Vasectomy is not effective immediately and other contraception must be used until two consecutive negative sperm counts are obtained.

COUNSELLING

This is a most important prerequisite of sterilisation.

1. The counsellor may be the family doctor who knows the couple, their children, their background and their home circumstances. He should also be experienced in contraceptive counselling, so that the request for sterilisation can be seen in relation to other methods of contraception. Not all general practitioners feel able to take on these functions and the family planning doctor may then be the best person to counsel the couple.

2. Counselling is an exchange of information. The couple should be seen together in a relaxed but professional atmosphere. They should provide details of themselves, their circumstances and the reason for their request, while the counsellor gives the facts about the operation, its risks and implications. On the basis of this exchanged information and, after a full discussion of the couple's attitudes, fears and possible misgivings, a decision can usually be made fairly easily.

3. Information from the couple.

    a. Relevant medical histories.

    b. Ages, occupations, social circumstances.

    c. Numbers, ages and health of their children; miscarriages; therapeutic abortions or ectopic pregnancies.

    d. Intended family size at the start of the relationship and the number of pregnancies subsequently unplanned.

    e. Previous and current contraception and any problems encountered or fears related to them.

    f. Duration and stability of the relationship and whether either partner has been married before.

    g. The quality of their sexual life.

4. Information from the counsellor.
a. A simple description of the techniques of both female and male sterilisation.
b. The possible risks, *failure rates* and side effects (Ch. 14).
c. Long-term sequelae.
d. Prospects of reversal.

Points for discussion

1. The implications of breakdown of the marriage, death of one partner, or of their children.
2. The myths which are associated with sterilisation, e.g. menstruation ceases, libido decreases, the menopause is more troublesome, women get fat or even that the operation reverses itself after seven years. Inaccurate information abounds and it is important that the couple should be able to discuss such matters with the counsellor.
3. Which partner is to be sterilised? This is normally a decision for the couple. However, concurrent disease in one partner may make the operation more hazardous for him or her or contraindicate it and the doctor will have to explain how this should influence their decision.

In the woman, a history of menorrhagia, even though it is currently controlled by the pill, or the presence of fibroids, may mean that hysterectomy would be indicated in the future and this should be taken into account. Also, a past history of pelvic or abdominal surgery or infection in the woman or scrotal surgery in the man may mean that the operation may be technically more difficult.

When a decision is made as to which partner is to be sterilised further details of the procedure can then be given.

Consent

Informed consent to operation is obtained from the individual and recorded. Although the consent of the spouse is not legally required, it is good practice to obtain it to ensure agreement between the couple (Ch. 14).

CLINICAL MANAGEMENT

Male sterilisation

Examination

Before a final decision is made the man should be examined. Previous herniorrhaphy, the presence of a hernia, hydrocoele or undescended testis and gross obesity often make vasectomy under local analgesia difficult. General anaesthesia may be indicated in these cases, or for the unduly nervous patient.

The woman should also be examined to exclude any condition which might require treatment that would render her sterile or would allow sterilisation to be carried out simultaneously and consequently make vasectomy unnecessary. Usually a pelvic examination and cervical smear should be carried out, unless they have been done recently and were normal.

Timing of operation

Vasectomy may be carried out at any convenient time. Sometimes the operation may be requested when the partner is pregnant so as to ensure a minimal period of male fertility postpartum. There is no contraindication to this, especially if the pregnancy was unplanned.

Preoperative advice

1. The scrotal area should be shaved before attending for operation.

2. A scrotal support or a pair of firm Y-front underpants should be brought to wear afterwards.

3. If possible the man should be accompanied home after the operation.

4. There is no need to avoid food or drink if local analgesia is used.

Techniques

General or local anaesthesia may be used.

A variety of techniques may be employed but the underlying principle is the same. The vas is palpated in the upper scrotum and held between the thumb behind and the forefinger and middle finger on each side in front. A small quantity of local anaesthetic is injected into the skin, vas and surrounding tissue for about 2 cm. A skin incision of 1 cm is made over the vas which is then exteriorised in its fascial covering. The fascia is incised longitudinally to prevent accidental severance of the vas which is then itself withdrawn from the fascia, divided and its ends ligated with catgut (Figs 10.9 and 10.10). One end is replaced within the fascial envelope, the other is left outside and the fascia is closed. Absorbable sutures are inserted into the skin incision and the procedure repeated on the other side. This technique

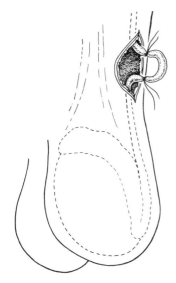

Fig. 10.9 Vas ligated.

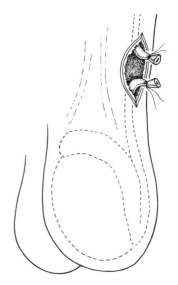

Fig. 10.10 Vas divided.

reduces the risk of subsequent failure since the cut ends are in different fascial planes. The potential for reversal is retained.

Alternative methods include:

1. Excising a small segment of vas. The excised portions, preserved in formalin, may be sent for histological examination at the time or kept and examined at a later date should semen analysis not become negative.

2. Looping each end back on itself.

3. Cauterising the cut ends.

Postoperative advice

A typed summary of postoperative advice and an address and telephone number to contact if complications arise is helpful (p. 222). The patient should also be given the following information verbally:

1. Skin sutures will dissolve spontaneously within approximately one week of operation.

2. Some scrotal swelling and possibly slight bruising is to be expected.

3. It is advisable to wear a good scrotal support, or firm Y-front underpants, night and day for the first two weeks.

4. Many men are able to go back to work the next day but it is preferable to avoid heavy manual work for two or three days.

5. Sexual activity may be resumed as soon as there is no further discomfort.

6. Since the man is not yet sterile alternative contraception must be used until two consecutive specimens of semen are sperm-free.

7. Seminal fluid will be examined after 12 weeks and 16 weeks and, if sperm are still present, at intervals thereafter. Instructions for the collection and delivery of specimens to the laboratory should be followed carefully.

Follow-up

1. It has been shown that between 24 and 36 ejaculations are required to clear the semen of sperm. Seminal analysis should be performed at 12 and 16 weeks after vasectomy. Ideally the results should be given to the patient in person. This avoids communication failure and also provides an opportunity to elicit and deal with any problems. In practice, however, this is often impracticable and results should then be sent by letter.

2. When two consecutive sperm counts are negative, both the patient and his general practitioner should be informed. Other contraception may then be abandoned. Thereafter no further follow-up is necessary.

3. In a small proportion of cases sperm persist in the seminal fluid for many months. Seminal analyses should be repeated at monthly intervals for six to nine months. If they remain positive at the end of that time the patient should be referred for further investigation. This normally entails exploration under general anaesthesia to check that reanastomosis (normally due to sperm granuloma) has not taken place or, rarely, that a third vas has not been missed.

Complications and their management

Short-term complications

1. Scrotal haematoma is not uncommon but normally requires no treatment beyond analgesia and local support.

2. Bleeding from the wound may be controlled by rest or by pulling the wound edges together with strapping, but if it persists restitching may be required.

3. Wound infection or epididymitis need antibiotic therapy, e.g. ampicillin 500 mg 4-times daily for 7 days.

Long-term complications

1. Psychological sequelae to vasectomy are kept to a minimum by adequate counselling and explanation before the operation. However, it has been shown that men with pre-existing sexual dysfunction may be adversely affected. In general, those who complain of an adverse effect on sexual performance after vasectomy nearly always have had a history of similar troubles before surgery.

2. Long-term physical complications have from time to time been suggested but none have so far been substantiated. There is no evidence of an increased risk of cardiovascular disease in men after vasectomy (Petitti et al 1982), although there is evidence that some vasectomised monkeys are at increased risk of developing atheromatous changes in the arteries.

3. Sperm antibodies may be produced which are only of relevance if reversal is contemplated because their presence may influence the man's fertility.

Female sterilisation

Examination

In addition to the routine screening procedures and pelvic examination, factors which might contraindicate or complicate the operation, such as previous abdominal operations and gross obesity, should be noted.

Timing of operation

1. Female sterilisation should be arranged during menstruation or the first half of the menstrual cycle to avoid an unsuspected pregnancy. This, however, may not always be practical.

2. Ideally, sterilisation of either partner should not be done in association with a pregnancy or in the puerperium. A decision taken during or immediately after pregnancy may be regretted at a later stage.

3. Female sterilisation performed in the puerperium or at termination of pregnancy is associated with a higher failure rate and increased risk of postoperative complications. However, in selected situations, particularly at the time of caesarean section, it is acceptable. The main proviso is that the decision to be sterilised should have been taken in principle before or during the pregnancy rather than during or just after labour.

Preoperative advice

1. If the woman is being admitted on the day of operation she should fast for six hours beforehand.

2. Preoperative shaving is not required.

3. If she is taking the oral contraceptive pill she should continue to do so right up to the time of admission.

Techniques

General anaesthesia is normally used, although spinal or local analgesia may be employed.

Laparoscopy (Fig. 10.11).

A pneumoperitoneum is created by the insufflation of carbon dioxide or nitrous oxide into the peritoneal cavity. Through a small subumbilical incision a trocar and cannula are introduced into the gas-filled space and the trocar replaced by the laparoscope. With the fibre-optic light source connected the pelvic organs are inspected. The operating forceps are introduced through a second cannula inserted in

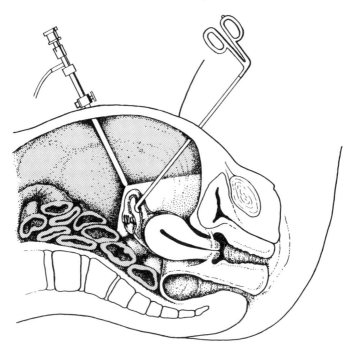

Fig. 10.11 Laparoscopy.

one iliac fossa or in the lower midline. Sterilisation is performed either by diathermy or the application of clips or rings (pp. 206, 207). After the release of gas from the peritoneal cavity the instruments are withdrawn and the tiny skin incisions closed with stitches of absorbable material or skin clips.

The patient normally returns home after 24 hours but in some cases may be fit for discharge the same day.

Minilaparotomy.

This is performed through a small suprapubic transverse or midline incision. The tubes are manipulated towards the incision in turn using a uterine elevator inserted vaginally and the sterilisation completed either by tubal ligation or occlusive methods similar to those used with the laparoscope.

Depending on the size of the incision and the individual patient's reaction most patients may be discharged from hospital within one to two days.

This method is to be preferred if the operation is done in the puerperium and when laparoscopy is made difficult by complicating factors, such as intra-abdominal adhesions, obscuring the view.

Postoperative advice

Most women find a leaflet containing a summary of the following information helpful and reassuring:

1. The small laparoscopic wounds are usually closed with absorbable sutures and require no further treatment. They will heal within 10 days. If skin clips are used they will be removed before leaving hospital or arrangements made to have this done at home.

2. Slight bruising and associated discomfort may sometimes be experienced round the wounds for a few days longer.

3. Gas remaining in the peritoneal cavity often causes some abdominal discomfort or shoulder pain for 24 hours or longer.

4. Light work may be resumed within two days and full activity within one week.

5. A minilaparotomy wound takes a few days longer to heal and heavy lifting should be avoided for about three weeks.

6. Sexual activity may be resumed when she feels like it and no contraception is required.

Follow-up

1. The patient should be seen after six weeks either in hospital, by her general practitioner, or by the family planning doctor.

2. Her menstrual history should be checked, and the abdominal wounds inspected. It is not always necessary to carry out a pelvic examination but this is indicated if she reports gynaecological problems or if there is any suspicion that she may be pregnant.

The first menstrual period after the operation may be delayed but this is rarely due to a pregnancy, although the possibility should be borne in mind, particularly if sterilisation was carried out in the luteal phase of the cycle. A delayed period is much more likely to be the result of stopping the oral contraceptive pill.

Complications and their management

Short-term complications.

1. Haemorrhage and infection are uncommon.

2. Thromboembolic disease is fortunately rare, but is more likely if the procedure is done at the time of caesarean section or in the puerperium.

3. Pain may be due to tubal ischaemia, particularly after the application of Falope rings, and may require analgesics for several days.

4. Bowel damage may occur at the time of operation if diathermy is used but is a rare occurrence. Early recognition of this complication is essential. It should be suspected in any patient with unexplained pain, pyrexia and abdominal rigidity occurring within the first two weeks. Such cases should be referred to hospital immediately.

5. The operation carries a small operative mortality — 10.4 per 100 000 operations (Chamberlain & Brown 1978).

Long-term complications.

1. There are no physical changes in women as a result of sterilisation. Cyclical ovarian activity and menstruation continue unchanged.

2. It has been suggested that more women suffer from menorrhagia after sterilisation with a consequent increase in the incidence of hysterectomy (Templeton & Cole 1982). It has been postulated that this may be due to interference with vasculature in the broad ligament, particularly after laparoscopic diathermy. There seems little justification for this suggestion. Women who have been sterilised are still subject to normal gynaecological diseases and all too often the sterilising operation gets the blame (Destefano et al 1983). This may occasionally be related to emotional attitudes which were perhaps not recognised during counselling.

3. Psychological problems are rare and normally occur in those in

whom problems existed in some form before the operation or in those who have had inadequate or unsympathetic counselling.

4. Bowel obstruction from adhesions is a rare complication.

5. A significant proportion of pregnancies after sterilisation are ectopic. This possibility should always be borne in mind.

REVERSIBILITY

A sterilising operation should primarily be designed to be as effective as possible with the minimum of risks rather than with a view to reversibility. However, despite extensive counselling it is inevitable that a few couples will request reversal of their sterilisation.

Those methods of female sterilisation which involve the least tubal destruction (clips, rings and some tubal ligation techniques) potentially offer the best prospects for reversal. Coagulation of the tubes usually results in a fairly extensive area of tubal destruction and makes reversal more difficult. Reversal involves open laparotomy, excision of the damaged portion of tube and then reanastomosis or reimplantation. The results of these operations are being improved by the use of microsurgical techniques and in competent hands 50–60% of women have had their fertility restored. In a small proportion of cases the pregnancy will implant in the fallopian tubes (Gomel 1980).

Reversal of vasectomy is technically feasible in many cases but the pregnancy rate is disappointingly low, possibly due to the formation of sperm antibodies. Sperm can be found in the ejaculate in about two-thirds of cases but pregnancy is achieved in only about one-third.

RISKS/BENEFITS

Although sterilisation operations do carry small but significant risks the overall benefits in terms of effectiveness and convenience tend to outweigh these, in the well-motivated and adequately counselled couple.

In a couple who have completed their family the procedure avoids the need for continued motivation in contraceptive usage and the long-term side effects of effective reversible methods such as the combined pill and IUD.

REFERENCES

Chamberlain G, Brown J C (eds) 1978 Gynaecological laparoscopy: working party report. Royal College of Obstetricians and Gynaecologists, London

Destefano F, Huezo C M, Peterson H B, Rubin G L, Layde P M, Ory H W 1983 Menstrual changes after tubal sterilisation. Obstet. Gynecol. 62: 673

Gomel V 1980 Microsurgical reversal of female sterilisation: a reappraisal. Fertility and Sterility 33: 587–597

Petitti D B, Klein R, Kipp H et al 1982 Physiologic measures in men with and without vasectomies. Fertility and Sterility 37: 438–440

Templeton A A, Cole S 1982 Hysterectomy following sterilisation. British Journal of Obstetrics and Gynaecology 89: 845–48

Vessey M, Lawless M, Yeates D 1982 Efficacy of different contraceptive methods. Lancet i: 841–842

Appendix

ADVICE FOR MEN WHO HAVE HAD A VASECTOMY

1. You have had a vasectomy under local anaesthetic. There is a slight risk of bleeding and swelling but this will be less likely to happen if you take the following advice.
   a. Take things quietly for the next 24 hours. Don't take strenuous exercise or lift anything heavy for 48 hours.
   b. Don't bath or shower or remove the dressing for 24 hours.
2. Care of the wound(s) — keep the dressing(s) covering the wound(s) in place with firm underpants or other scrotal support. You may find it comfortable to continue wearing support for a few days and even during the night.
   24 hours after the operation have a bath or shower and soak off the dressings. Bath or shower every day until the wound is healed.
   Dry the area of the wound(s) carefully and no further dressing(s) should be required.
   If there are stitches in the skin they will fall out themselves in two to three weeks.
3. Pain may occur at the site of the operation when the local anaesthetic wears off. A mild analgesic (for example soluble aspirin 4-hourly) usually helps.
4. A certain amount of bruising may occur and clear discharge may come from the wound for a few days. If excessive bleeding, pain, bruising or swelling occurs go to bed and get in touch with your own general practitioner who will advise you whether anything needs to be done.
5. You may resume intercourse when you feel like it but it is important to continue to use other contraceptives until we tell you that the operation is complete. The operation is not considered complete until there are no sperm in the ejaculate. It is important to have regular ejaculations otherwise sperms may remain in the tubes in large numbers for a very long time.

6. Laboratory tests — vasectomy is regarded as complete when two consecutive specimens of seminal fluid are shown to be free from sperm. You have been given two containers in which to send your specimens to the laboratory. Make sure that the form accompanying this specimen is filled in accurately and the date recorded.
   a. The first specimen is examined 12 weeks after the operation and the second specimen 3 weeks later. Your first specimen should be sent to the laboratory in the week beginning.

   (Date) ....
   b. Seminal fluid may be collected directly into the container either by masturbation, or by withdrawing just before ejaculation takes place. When a sheath is used during intercourse, the whole of the specimen of seminal fluid is required and should be emptied into the container. The sheath should not be included.
   c. If sperm are found in either of the two specimens of seminal fluid you will be asked to send further specimens to the laboratory at 3-weekly intervals until 2 consecutive specimens are negative. Instructions on how to do this along with the necessary forms and containers will be sent to you if necessary.
7. As soon as the operation can be regarded as complete you and your general practitioner will be informed. Please do not phone the clinic for results.
8. Vasectomy is one of the most reliable methods of contraception, but very occasionally the operation has to be redone because the cut ends of the tubes (vasa deferentia) grow together again, and fertility is restored. We would like to stress that the chance of this happening is very small indeed. Nevertheless if, in the future, your partner should miss a period she should report this to the clinic or to her own doctor without delay.
9. Address and telephone number to contact in an emergency.

Postcoital contraception

A variety of methods of postcoital contraception (PCC) have been advocated, such as postcoital douching and vigorous physical exercises including unusual body gyrations immediately after intercourse in an attempt to dispel semen from the vagina. Before occlusive contraceptives became available postcoital douching was widely practised. Solutions containing a variety of substances such as soap, lemon juice or vinegar and latterly even carbonated beverages were popular but unfortunately not effective. Since it has been shown that sperm will enter cervical mucus within 90 seconds of ejaculation the poor performance of postcoital douching is readily explained.

Despite the general availability of effective contraception there remain occasions when unprotected intercourse takes place, or accidents happen with barrier methods at the fertile time of the cycle. The woman may then wait and do nothing or she may ask for PCC.

Although not every single act of intercourse around midcycle necessarily results in pregnancy, the risk is 30% or more depending on the inherent fertility of the couple (Tietze 1960).

Hormones are most commonly used as postcoital agents but IUDs are also effective. Many other compounds, e.g. danazol, antioestrogens and antiprogestogens work in experimental animals but are not generally available to women.

Whichever method is chosen it is important that it (1) should be as free as possible from side effects, (2) should not affect cycle control or endogenous hormone production, (3) should be free from teratogenic side effects (in case of failure).

Patients should understand that postcoital contraception is an emergency measure only and is not suitable as a regular method of contraception.

METHODS

Hormonal methods

Commonly called 'morning after' pills. The CSM gave approval for the use of hormones as postcoital contraceptive agents in January 1984. In their view the combined oestrogen/progestogen regimen was the method of choice.

Combined oestrogen/progestogen preparations

Dose: 100 μg ethinyloestradiol + 500 μg levonorgestrel (i.e. 2 tablets of Ovran or Eugynon 50) given immediately and repeated after 12 hours.

This regimen has proved very effective (Yuzpe et al 1982) and is the way in which hormones are now most commonly used as PPCs. Schering PC4 is marketed specifically as a postcoital contraceptive.

Oestrogens

These were the first compounds to be tried. The results of the relevant clinical studies are set out in the references (Dixon et al 1980, Haspels 1976).

1. Diethylstilboestrol

Dose: 50 mg daily for 5 days, usually given as 25 mg twice a day. The administration of this preparation to women in pregnancy has been shown to be associated with the occurrence of the rare clear-celled carcinoma of the vagina and its predisposing adenosis in female offspring in their teenage years. For this reason it is not recommended.

2. Ethinyloestradiol

Dose: 5 mg daily in divided doses for 5 days. (2 × 1 mg tablets in the morning: 3 × 1 mg tablets in the evening).

3. Premarin (conjugated oestrogen)

Dose: 30 mg daily in divided doses for 5 days.

Oestrogens alone are not now used so often as PCCs. When they are it is advisable to prescribe an antiemetic such as Ancoloxin prophylactically to try to ensure completion of the full course of treatment.

Progestogens

Several progestogens have been tried as postcoital agents, used routinely after each act of intercourse.

1. Levonorgestrel

Dose: 400 μg within 3 hours of intercourse.

2. Quingestanol acetate

Dose: 0.5 mg postcoitally.

Progestogens used in this way have a high failure rate and are not recommended for PCC.

Mode of action

Postcoital hormone therapy appears to act in several ways:

1. The endometrium is rendered unfavourable for nidation. There is desynchronisation of the delicate sequential process, with the glands lagging two to six days behind the changes in the stroma. There is also evidence that the oestradiol and progesterone receptors may be altered.

2. Interference with normal corpus luteum function is suggested by a reduction in plasma progesterone levels, an absence of the luteal elevation of basal body temperature and a shortening of the luteal phase. This correlates with the earlier onset of the period following therapy.

3. Ovulation is usually inhibited if treatment is commenced early enough in the cycle, i.e. before the preovulatory surge in oestrogen. The positive feedback of the oestrogen rise is inhibited and often the luteinising hormone peak is abolished. However, if administration of the hormones follows ovulation this does not occur.

4. There may also be interference with ovum transport and direct action on the blastocyst.

Intrauterine devices

Inert devices such as the Lippes Loop or copper releasing devices such as Copper 7, Copper T, Multiload, Novagard or Nova T may be inserted postcoitally.

Mode of action

The presence of an IUD primarily prevents implantation by changes in the endometrium induced both by the IUD itself and by an action of the copper ions which affects the biochemical processes and enzymes involved in implantation. The copper ions affect the endometrial secretions, inhibit their synthesis and also stimulate local prostaglandin production. All these activities interfere with the mechanism of nidation of the blastocyst into the endometrium.

Menstrual regulation

Although this technique is sometimes listed with postcoital contraception it is really early abortion (p. 240).

EFFECTIVENESS

1. Postcoital oestrogens have been used to treat many thousands of patients and only a few pregnancies have occurred. These were usually associated with:
    a. Too low a dose of oestrogen.
    b. Failure to complete the course of treatment.
    c. Oestrogen therapy started too late.
    d. Multiple exposure to risk before and after treatment.
The failure rate appears to be less than 1.0% with a 10% incidence of ectopic pregnancy in those for whom the method fails.

2. Oestrogen/progestogen preparations are highly effective. In the multicentre Canadian studies (Yuzpe et al 1982) the failure rate was 1–2% (1.0% when the pills were taken correctly). A higher pregnancy rate of 5% for midcycle exposure (Tully 1983) suggests that this method may not always be as effective as the oestrogen regimen described above. It is, however, more acceptable to patients because it produces fewer side effects.

3. In one study where a copper IUD was fitted postcoitally in 299 women after unprotected intercourse around the time of ovulation no pregnancies occurred (Lippes et al 1979).

INDICATIONS

Postcoital contraception is a 'first aid' measure and not one to be used routinely.

The common indications are:

1. Isolated unprotected intercourse.

2. Failure of a barrier method — burst sheath, sheath falling off, dislodgement of a cap, etc.

3. Complete or partial expulsion of an IUD.

4. Fear of pregnancy following rape.

CONTRAINDICATIONS

1. Pregnancy.

2. Multiple exposure. Higher failure rates occur when hormonal methods are used in cycles where unprotected intercourse occurred on more than one occasion.

3. A clinical history of contraindications to oestrogen therapy, e.g. thromboembolism. Although this complication has not been reported despite relatively high doses of oestrogen with postcoital contraceptive therapy, an IUD should be considered instead.

4. A history of recent pelvic inflammatory disease. Combined oestrogen and progestogen therapy should be used in preference to an IUD.

5. A history of ectopic pregnancy is considered by some to be a contraindication to any type of PCC. Others are not convinced that PCC will increase the risk of another ectopic. Combined oestrogen/progestogen therapy is preferable to IUD insertion.

6. Postcoital contraception is not advocated for women who have missed one or more contraceptive pills. They should be advised to follow the standard instructions for missed pills (p.80).

ADVANTAGES

The advantages outweigh the disadvantages.

1. The treatment is effective. Pregnancy is rare following postcoital therapy, particularly with the IUD.

2. Hormonal treatment is short — 12 hours to 5 days depending on the method.

DISADVANTAGES

Hormonal methods

1. Need to start within 72 hours of intercourse.

2. Further coitus following therapy must be avoided until the

patient is established on a reliable method of contraception otherwise the method may fail.

    3. Side effects:

      a. With oestrogen therapy nausea occurs in up to 70% of patients and vomiting in about 35%. Generally these symptoms are not severe enough to stop treatment and may be controlled with an antiemetic.

      b. With oestrogen/progestogen therapy nausea occurs in 50–60% of patients and vomiting in 20–30% — usually after the second dose. Symptoms cease within 72 hours.

    4. The period following therapy may come early or be delayed.

Intrauterine devices

    1. The usual contraindications to the use of an IUD apply.

    2. The risk of pelvic inflammatory disease, particularly in young women who frequently change partners, may make this method inappropriate for them.

    3. IUD insertion may be difficult or painful particularly in the nulliparous patient. This has to be balanced against the systemic side effects associated with hormonal methods.

CLINICAL MANAGEMENT

Ideally the provision of postcoital contraception should be limited to subjects presenting within 72 hours of a single midcycle exposure to pregnancy. This time can be extended up to 5 days if an IUD is used, provided insertion is carried out before the 20th day of a 28-day cycle, i.e. prior to nidation.

    It is essential that an accurate record of the consultation, written at the time and dated be kept.

Assessment

History

It is important to take a careful history and record:

    1. Age and parity.

    2. Relevant medical, surgical and reproductive history.

    3. Date of last menstrual period.

    4. Details of patient's normal menstrual cycle.

    5. Calculated date of ovulation.

6. Day(s) in the cycle when unprotected intercourse took place.

7. Number of hours since first episode of unprotected intercourse.

8. Contraceptive being used, if any.

Examination

1. Carry out a pelvic examination especially
   a. if there is any suspicion that the patient may have been pregnant in preceding cycle(s);
   b. to check the size and position of the uterus and the normality of the adnexa to assess suitability for an IUD.

2. Record the blood pressure. Older women (over 35) receiving high-dose oestrogens should have their blood pressure checked.

Counselling

1. Explain the methods available, their risks, failure rates and the importance of follow-up. Discuss the possibility that the next menstrual period may come earlier or later.

2. Explore the patient's attitude to possible failure of the regimen and continuance of the pregnancy. Many patients fear that if hormone therapy is employed and fails there will be a risk of fetal damage. So far there have been no reported fetal abnormalities following failure of hormonal PCC so the risk is largely theoretical. However, one cannot reassure the patient completely on this point since there have been insufficient numbers of pregnancies proceeding to term to evaluate the risk adequately. It is often helpful to explain that many women have become pregnant while taking the oral contraceptive pill and the risk of detrimental effect on the baby has been found to be very small.

3. Make the final decision about whether to use postcoital contraception only after full discussion with the woman.

4. Discuss future contraception.

Choice of method

This will depend on the interval between intercourse and the time when the patient is seen.

1. Midcycle 'risk', i.e. intercourse occurring at the time of ovulation ± 3 days (days 11–17 in a 28-day cycle). Hormonal methods are best if taken within 72 hours by women with regular cycles.

2. Postovulatory 'risk', i.e. later than day 17 or 18 of a 28-day cycle, or more than 72 hours after a risk. IUD insertion is the method of choice.

3. Early preovulatory 'risk', i.e. before day 10 in a 28-day cycle. Either hormones or an IUD may be used. If hormones are prescribed ovulation may be delayed for 7–10 days. In this instance it is particularly important to ensure that ·the patient does not have intercourse later in the cycle without adequate contraception.

Instructions to patients

Hormonal methods

1. Tell her exactly when to take the pills. She can take the first dose before she leaves the clinic or surgery.

2. If she vomits she should take another 1 or 2 pills, depending on the regimen used. Provide her with a sufficient supply to cover this eventuality.

3. If she forgets or vomits the second dose of combined oestrogen/progestogen pills, provided she is still within the 72-hour interval she should restart the whole regimen, i.e. she should take two pills immediately and a further two pills after 12 hours. Alternatively advise her to come back to discuss IUD insertion.

4. Abstinence or a barrier method should be used until the onset of the next menstrual period.

5. Warn her that her next period may come early or be delayed.

6. Stress the importance of attending for follow-up, *whether or not she has had a period or scanty vaginal bleeding.*

Intrauterine devices

1. Explain about postinsertion cramps, how to relieve them with analgesics and about postinsertion spotting.

2. Stress the importance of follow-up.

Follow-up

All patients should be seen at least once following medication, preferably at the time of the expected next menstrual period or within the subsequent five days. The patient must be instructed to return whether she menstruates or not. It is important to ensure that:

1. She is not pregnant,

2. She has effective contraception — this should have been discussed already during counselling, particularly if she wants to take the pill (see below),

3. The question of termination of pregnancy is discussed if the method fails.

Patients who have a normal 'period'

In these cases postcoital contraception has been effective. At this visit it is important to establish adequate long-term contraception. Those who have had an IUD inserted may wish to keep it. If the patient wishes to take the combined pill, a preparation which she starts on the 5th (not the 1st) day of the period should be prescribed. Pills which must be started on the 1st day of the period are best avoided, since it is impossible to predict whether a 'normal' period is starting on the first day of bleeding.

Patients who have scanty bleeding

Pregnancy must be excluded.

Patients who do not menstruate

The period may have been delayed by hormone therapy or the patient may be pregnant. If she is pregnant she may wish to consider termination (Ch 12). If an IUD has been inserted and pregnancy is confirmed, should she wish to continue with the pregnancy the IUD should be gently removed if the tail is still visible.

The risks of postcoital contraception are minimal and are greatly outweighed by the benefits conferred by preventing unwanted pregnancy.

Patients should have easy access to such help and advice. Many general practitioners and family planning clinics now offer this service. Its availability should be widely publicised. Some clinics provide telephone numbers which patients may ring for help. It is also important that the staff in the clinic or general practice know that this service is being offered so that they can ensure that the patient sees the doctor quickly by giving her an early appointment.

REFERENCES

Dixon G W, Schlesselman J J, Ory H W, Blye R P 1980 Ethinyloestradiol and conjugated estrogens as postcoital contraceptives. Journal of the American Medical Association 244: 1336–1339.

Haspels A A 1976 Interception: postcoital estrogens in 3016 women. Contraception 14: 375–381.

Lippes J, Tatum H J, Malik T, Zielezny M 1979 A continuation of the study of postcoital IUD. Family Planning Perspective II: 195.

Tietze C 1960 Probability of pregnancy resulting from a single unprotected coitus. Fertility and Sterility 11: 485–488.

Tully B 1983 Postcoital contraception — a study. British Journal of Family Planning 8: 119–124.

Yuzpe A A, Percival Smith R P, Rademaker A W 1982 A multicenter clinical investigation employing ethinylestradiol combined with dl-norgestrel as a postcoital contraceptive agent. Fertility and Sterility 37: 508–513.

Termination of pregnancy

Although not strictly a method of contraception, termination of pregnancy (TOP) is an important and necessary back-up facility in family planning services. Forty percent of the world's population has access to abortion more or less on demand and a further 26% is allowed it for social or medical reasons. In Britain more than 160 000 legal TOPs are carried out each year. Illegally induced abortions with their tragic sequelae have all but disappeared. After liberalisation of the law in any country the number of abortions carried out usually rises rapidly but there is evidence that where good family planning services are available the number always levels off and may even decline.

It is recommended that TOP is not carried out after 20 weeks gestation other than in exceptional circumstances, e.g. for genetic reasons, severe fetal abnormality or severe maternal disease.

INDICATIONS

The present law permits TOP to be carried out under the terms of the Abortion Act (1967) (p. 265).

1. 'that the continuance of the pregnancy would involve risk to the life of the pregnant woman ... greater than if the pregnancy were terminated.'

Interpretation: the spirit of this clause is that if a woman is seriously ill, e.g. with severe heart or kidney disease, TOP would be less dangerous than continuing the pregnancy to full term. The operation is therefore to save or preserve life. However, under the law, this clause may be interpreted to mean that below 12 weeks gestation the risk of termination is substantially less than the risk of full-term pregnancy. It could, therefore, be argued that it would be safer to terminate that pregnancy than to allow it to go to term. Any abortion under 12 weeks could thus be interpreted as being legal.

2. 'that the continuance of the pregnancy would involve risk ... of injury to the physical or mental health of the pregnant woman greater than if the pregnancy were terminated.'

Interpretation: this clause is very difficult to interpret precisely. It relates particularly to the patient who, for various reasons, cannot face continuing with the pregnancy and develops a reactive depression. This may proceed to mental or physical deterioration and if abortion were considered likely to reverse this process, it is clearly justified.

3. 'that the continuance of the pregnancy would involve risk ... or injury to the physical or mental health of ... any existing children of the family greater than if the pregnancy were terminated.'

Interpretation: if another baby would substantially affect the mental or physical health of the existing child or children then TOP would be legal.

4. 'that there is a substantial risk that if the child were born it would suffer from such physical or mental abnormality as to be seriously handicapped.'

Interpretation: this covers cases of possible fetal abnormality, e.g. following maternal rubella or where congenital abnormality such as Down's syndrome, spina bifida or sex-linked disease has been identified.

CONTRAINDICATIONS

There seem to be no definite contraindications to abortion other than those that cannot be placed under any of the clauses of the Abortion Act. However, special consideration should be exercised:

1. When a patient shows signs of ambivalence concerning her decision.

2. When a patient appears to be being pressurised into having a termination by her consort or parents.

Further counselling, especially by an experienced counsellor, should be available as termination may be potentially harmful to some of these patients.

Abortion should not be used as an alternative to contraception and every effort should be made to discourage such practice.

CLINICAL MANAGEMENT BEFORE TERMINATION OF PREGNANCY

Counselling

A woman with an unwanted pregnancy is often in a dilemma. To continue with the pregnancy may present insoluble problems and seem quite impractical whereas to have an abortion may seem equally abhorrent. She may not act entirely objectively under the strain of her predicament and sympathetic counselling is often needed to help each woman to reach the right decision for her. On the other hand, by the time many women see the doctor they are so certain that they wish the pregnancy terminated that they reject offers of counselling and only want referral to the appropriate agency.

Counselling is a process of discussion between the woman and the counsellor. It should be informed, nondirective and should take place in a relaxed informal setting. It begins with the first professional contact — with the general practitioner, the clinic doctor or nurse. More detailed counselling is best carried out by someone who is trained in such techniques and who may be a doctor, nurse, social worker or lay person. It can be helpful to the patient to discuss the problem with someone who is not involved in making the final decision whether or not to accept her request for termination, as the doctor inevitably is.

Counselling help is available from the charitable pregnancy advisory services (the British Pregnancy Advisory Service and Pregnancy Advisory Service) which have over 30 centres throughout the country. Brook Advisory Centres and Marriage Counselling

Services also provide counselling for women with unwanted pregnancy, whereas the counselling offered by Lifeline is directed towards helping the woman to decide to continue with her pregnancy.

If the patient agrees, her partner or parents may be interviewed as well. This not only helps to build up a comprehensive background picture to the problem but enables the counsellor to assess whether undue pressure is being put on the woman to reach a decision in one way or the other.

Wherever possible, patients seeking TOP should not be seen along with obstetric, infertile or routine gynaecological patients in hospital.

The aim of counselling is to help the woman to:

1. Decide her best course of action.
2. Decide what her real wishes are.
3. Take responsibility for her own ultimate decision.
4. Avoid serious regrets about it later.
5. Understand how she came to be in her present difficulty so as to avoid finding herself in the same situation again (Simms 1973).

The woman should be given every opportunity to express her own fears and doubts and to explore her own feelings and attitudes. Many women welcome the opportunity just to talk about the predicament in which they find themselves. To do so often helps them to reach the best decision for them. Discussion can range over a wide variety of subjects depending on the circumstances of each individual case, e.g.:

Why is this pregnancy so unwanted?

Is this the only problem in the woman's life or is it just another to add to the difficulties she is already facing?

Is she concerned about what her family and friends will say if they find out?

Does she have a stable relationship?

Does she have problems with using contraception, etc?

The counsellor must also make sure that the woman is given as much information as possible to help her make the decision.

1. *The alternatives to abortion* — adoption, fostering, keeping the baby herself — and the support available to her if she pursues one of them. Information on maternity leave, pay, grants and other benefits is contained in leaflets which are available from the local social security offices.

2. *What having an abortion involves* — the way in which it will be carried out (appropriate to the gestation) the type of anaesthetic and the possible risks.

3. *How long she will probably have to stay in the hospital or nursing home.*

4. *The importance of reaching a decision without too much delay on the one hand, yet leaving adequate time to consider all the implications on the other.* Abortion carried out in the early weeks of pregnancy is a much simpler procedure associated with fewer risks and complications. However, the woman should understand that she is completely free to change her mind and decide against termination when she has had time to consider the matter further.

5. *The cost* — free under the National Health Service, or the approximate fee charged by the agency or consultant if she is referred privately.

6. *When her periods are likely to return.*

7. *Future fertility* — abortion carried out during the first trimester by modern techniques does not significantly impair future fertility, provided the procedure is uncomplicated and the woman does not undergo repeated terminations of pregnancy.

8. *Future contraception* — this should be discussed and where possible a firm decision made about it. An IUD can safely be inserted as soon as the uterus is evacuated. There is no increased risk of infection or excessive uterine bleeding. If the couple have had sufficient time to reach a decision about sterilisation it can be carried out at the time of surgical termination of pregnancy (Ch. 10).

Some of this information can be incorporated in a leaflet which can be given to each woman.

The following women need particularly careful counselling and, if the pregnancy is terminated, often require continuing support thereafter.

1. Young girls, whether they are married or not.

2. Women requesting late termination — the fact that they have delayed seeking help is often a sign of their ambivalence.

3. Those who are obviously distressed or who have a history of psychiatric problems.

4. Patients who have TOP carried out on medical or genetic grounds or because of the risk of fetal abnormality.

5. Women who have been raped.

Assessment

History

See Chapter 2. A careful gynaecological and social history is particularly important.

Examination

1. Pregnancy test: a pregnancy test is reliable 14 days after the first missed menstrual period. Arranging for the test, obtaining the result and discussing it with the patient should be achieved with as little delay as possible.

2. Vaginal examination — this should be carried out to confirm the presence of an intrauterine pregnancy and assess the gestation.

3. Ultrasound scan may be indicated to confirm gestation or if an ectopic pregnancy is suspected.

Referral

Once a patient decides that she wishes to be considered for TOP she will have to be referred to a gynaecologist. Local facilities for and attitudes towards TOP differ greatly. One of the most important factors in determining whether or not termination will be carried out is the attitude of the consultant whom the patient sees.

The referring doctor should be conversant with the local situation and use that knowledge to ensure speedy, realistic and sympathetic consideration of each woman's case. Where the NHS cannot or does not provide this standard of service, patients should be referred to one of the charitable organisations or to a consultant privately.

If a doctor is unwilling to accede to a patient's request for abortion on moral, ethical or religious grounds he should advise her that she is within her rights to seek a second opinion from another doctor and that this should be done with the minimum of delay.

Consent to operation

See Chapter 14.

TECHNIQUES OF TERMINATION OF PREGNANCY BY DURATION OF PREGNANCY

The techniques employed will depend on a number of factors the most important being the duration of the pregnancy and the preference of the gynaecologist carrying out the operation. They are described in three main groups.

Very early Termination of Pregnancy

This is TOP carried out up to seven weeks after the last menstrual period and is also known as menstrual regulation, menstrual extraction or menstrual induction.

Medical techniques

Prostaglandins (PG) both PGE_2 and $F_{2\alpha}$ have been used successfully, although side-effects such as nausea, vomiting and diarrhoea are a problem. Newer analogues of PG which have a lower incidence of gastro-intestinal side effects may be preferable but are still in the clinical trial stage. Efficacy of these methods is still only 90–95%

Surgical techniques

The uterus is evacuated through a Karman curette by suction from a syringe with a lock (Fig.12.1) or from a small vacuum pump. The technique can be carried out under local anaesthetic infiltration of the cervix or basal narcosis with diazepam. It is rapid and effective but has side effects similar to suction TOP.

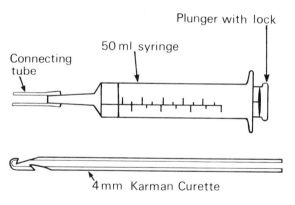

Fig. 12.1 Menstrual extraction kit.

First trimester termination of pregnancy

TOP up to the 13th week of gestation is very similar to menstrual extraction but dilatation of the cervix is more frequently needed and a suction pump is always required as a vacuum source.

Anaesthesia

General: although general anaesthesia is emotionally kinder to the patient than local analgesia it does subject her to small but definite risks. She also takes longer to recover and return home is delayed.

Local: this is very suitable for pregnancies up to 12 weeks gestation and is becoming more popular. The following techniques may be used:

1. Paracervical or intracervical block with 1% lignocaine with adrenaline 1 in 200 000.

2. Intracervical block plus extra-amniotic instillation of local analgesic.

3. Multiple site cervical injections of local analgaesic to 'block' the nerve supply to the internal os and cervix.

Dilatation of the cervix prior to operation

This is required more often in primigravid than multigravid patients. Too much dilatation is neither necessary nor safe.

1. Using metal dilators, the cervix is gently dilated to the size of the appropriate curette. A general guide is that a curette 2 mm less than the number of weeks gestation should be employed, e.g. at 10 weeks gestation an 8 mm curette should be used.

2. Prostaglandins and their analogues may be used by the intramuscular, extra-amniotic or vaginal routes. Analogues administered as a vaginal pessary or by intramuscular injection appear to be best. They are very effective, have a significant effect on cervical dilatation within three to six hours and are easy to administer.

3. Laminaria tents inserted into the cervical canal the day before operation. These small cylinders of laminaria digitala or japonica, 6–8 mm by 2–4 mm in size, are intensely hygroscopic, swelling to 3–5 times their original diameter when placed in the cervical canal.

4. Magnesium sulphate osmotic dilators (Lamicel) are used in the same way as the laminaria tents and are very effective two hours after insertion. They consist of 0.5 g magnesium sulphate, which is very hygroscopic, incorporated into a rigid vehicle of polyvinyl alcohol foam.

Suction curettage

This is the commonest method of TOP. It is safe, simple and quick. A suitable catheter connected to suction apparatus is inserted into the uterus through the previously dilated cervical canal and the uterine contents evacuated. Most catheters are designed to allow curettage as well as suction. This achieves removal of the decidual lining and effective emptying of the uterus.

Suction catheters (Fig. 12.2) The original rigid metal catheters have now been replaced by flexible plastic models of the Karman type. A rigid plastic catheter — the Berkley curette — is also very effective and is widely used.

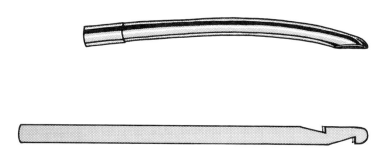

Fig. 12.2 Flexible Karman curette (below). Rigid Berkley curette (above).

Suction appliances. A suction appliance capable of achieving high negative pressure of at least 600 mmHg at sea level with a 'fail safe' mechanism to ensure that only suction may be generated is essential. The tubing and piping should be of adequate dimension so that the products of conception do not clog the system. Products of conception should also be easily visible in the receptacle. The Berkley, Synevac and Matburn pumps, which are electrically driven, meet most of these requirements. Portable mechanical suction pumps are also available.

Evacuation of the uterus. Once adequate dilatation of the cervix has been achieved the appropriate suction curette is inserted into the uterine cavity. In cases of retroversion great care must be exercised to avoid perforation of the uterus. Negative pressure, as high as possible initially, is commenced. Clear plastic catheters and tubing allow inspection of the products of conception as they are evacuated. Liquor, membranes, fetal and placental tissue may be identified. As the uterus empties it decreases in size. In cases where there is doubt that evacuation is complete the uterine cavity can be explored with a sharp curette.

Once the uterus is empty further bleeding is minimal. Continued blood loss usually indicates retained products of conception but if the uterus definitely feels empty an oxytoxic should be given.

The products of conception should be inspected. If there is any doubt about the diagnosis of intrauterine pregnancy they should be sent for histological confirmation, otherwise an ectopic pregnancy may be missed.

Sharp curettage

The traditional method of evacuating the uterus with a sharp curette is no longer recommended.

Second trimester termination of pregnancy

TOP during the second trimester is more difficult, has a higher complication rate and may require subsequent curettage under general anaesthesia to complete uterine evacuation. The main methods in current use are as follows:

Intra-amniotic injection

1. *Hypertonic saline.* Local anaesthetic is injected into the skin and underlying structures. A Tuohy needle is introduced trans-abdominally into the amniotic fluid (Fig. 12.3) midway between the symphysis pubis and the fundus of the uterus in the midline. Two hundred millilitres of liquor are removed and 200 ml of 20 or 22.5% hypertonic saline slowly instilled. This technique has a number of disadvantages:

   a. The intra-amniotic space is difficult to enter before the 16th week of pregnancy.

   b. If the injection is misplaced, e.g. into the placenta or uterine muscle, rapid absorption occurs with fatal results.

   c. There is a long injection–abortion interval of 28–30 hours.

   d. Twenty five per cent of cases require subsequent evacuation of retained products.

2. *Forty per cent urea.* The procedure for injection is similar to that for saline. Although the injection–abortion interval is long — over 48 hours — complications are less and no serious side effects occur from rapid absorption.

3. *Prostaglandin E_2 or $F_{2\alpha}$* may be given as a single intra-amniotic injection (Fig. 12.3). The injection–abortion interval is 12–16 hours. Repeated injections are not recommended.

Intramuscular PG analogues can be used to complete abortion initiated by intra-amniotic prostaglandins.

Intra-amniotic injections

- Hypertonic saline (22½%)
- Urea
- Prostaglandins

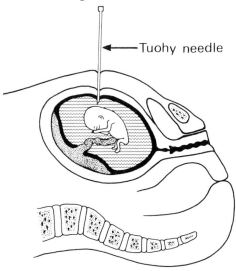

Fig. 12.3 Intra-amniotic injections.

4. *Combination therapy.* Intravenous syntocinon can be used along with intra-amniotic injections to decrease the induction–abortion interval.

Intra-amniotic urea can be combined with PG to decrease the induction–abortion interval.

Extra-amniotic instillation

Prostaglandins either in solution or in a gel (Tylose) instilled through the cervical canal into the extra-amniotic space are effective in inducing abortion (Fig. 12.4). This method avoids the risks of intra-amniotic injection, is easier to perform and has a lower incidence of side effects. Low doses (250 μg) of PGE_2, $F_{2\alpha}$, or their analogues are given every 4–6 hours until abortion occurs. The mean time till abortion is 12–16 hours. The method can be combined with extra-amniotic oxytocin administration.

Ravinol (an acridine dye) has also been used by the extra-amniotic route but is not as effective as PG.

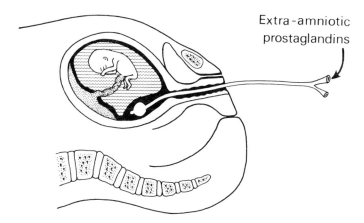

Fig. 12.4 Extra-amniotic prostaglandins.

Vaginal prostaglandins

When given as a pessary (either slow- or fast-release base), $PGF_{2\alpha}$, E_2, or their analogues will cause abortion. Repeated administration every 4–6 hours is necessary. The newer analogues have a low incidence of gastro-intestinal side effects. Efficacy is high and induction–abortion interval short — approximately 10–12 hours. Some patients require subsequent evacuation of retained products.

Other methods

1. *Hysterectomy.* This is now very rarely performed but is occasionally indicated.

2. *Dilatation and evacuation.* This method requires dilatation of the cervix to at least 12 mm, drainage of the liquor, removal of the placenta and crushing and removal of fetal parts. Great care must be exercised and it is not recommended except for experienced operators.

CLINICAL MANAGEMENT AFTER TERMINATION OF PREGNANCY

Early TOP is a simple, safe procedure. The patient recovers quickly and may go home the same day or the day after.

Later TOP may be associated with more serious complications which are outwith the scope of this book.

1. Tell the woman about the length of time for which she may expect bleeding or discharge to continue, how to recognise possible complications and whom to contact should they occur.

2. The Rhesus status having been established prior to TOP, anti-D immunoglobulin should be administered in an appropriate dose as soon as possible after operation in those who are Rh negative.

3. Contraception should be started (p. 35).

4. A follow-up appointment should be made.

Complications

Complications may arise at the time of TOP, shortly afterwards or at a later date.

At the time of TOP

1. *Uterine perforation.*

This may occur during dilatation of the cervix or when the curette is inserted into the uterus. If perforation is suspected at this early stage, laparoscopy should be performed and evacuation of the uterus completed under laparoscopic supervision.

If bowel is withdrawn this is a serious complication. A laparotomy should be performed and the rent in the uterus oversewn. The bowel should be inspected and the damaged area resected.

If a small perforation is suspected after the uterus is thought to be empty a conservative policy may be adopted and the patient kept under careful observation particularly with regard to pulse, blood pressure and vaginal bleeding.

2. *Vaginal bleeding.*

This may be due to uterine or cervical trauma or uterine atony. Oxytocics should be given and intravenous replacement therapy instituted. The cervix should be inspected and any tears sutured. If uterine damage is suspected laparoscopy or laparotomy will be necessary. Hysterectomy may be the only way to control haemorrhage.

3. *Anaesthetic complications if general anaesthesia is used.*

Within one week of TOP

The patient should be told to see her doctor if she develops heavy vaginal bleeding, lower abdominal cramps, pain or general malaise.

1. *Excessive vaginal bleeding.*

This is generally due to retained products of conception. Placental tissue may be left behind in 2–5% of all cases. If bleeding is not excessive ergometrine tablets 0.5 mg t.i.d. for 5 days may be effective in stopping it. If bleeding persists further exploration of the uterus is indicated and antibiotic cover may be used.

2. *Pelvic infection.*

Mild forms of pelvic infection will present with vaginal bleeding, lower abdominal pain or discomfort and may or may not be associated with general malaise. Mild cases generally respond to antibiotic therapy with metronidazole 200 mg t.i.d. and amoxycillin 500 mg t.i.d. for 7 days.

Severe cases of pelvic sepsis are not common but if they occur the patient should be referred to hospital immediately.

Late complications

1. *Guilt or regret.*

Many patients feel guilty, not so much about having had an abortion as about having become pregnant in the first place. Others truly regret their decision to terminate the pregnancy. Postabortion counselling should be available for those who require it.

2. *Long-term psychiatric sequelae.*

Serious psychiatric sequelae are very rare (Brewer et al 1977). They usually occur in women who have had previous psychiatric problems.

3. *Subsequent obstetrical problems*

a. Cervical incompetence — this may result from dilatation of the cervix either during suction evacuation of the uterus or prostaglandin induced termination (Schulz et al 1983). It predisposes to late spontaneous abortion and premature labour. The incidence of these complications in pregnancies following modern methods of TOP is believed to be much lower than with older techniques.

b. Subfertility — infection following either first or second trimester abortion may lead to tubal damage and consequent subfertility. In the past it was estimated that 2–5% of patients became infertile following TOP. The risk of subsequent infertility following termination of pregnancy carried out by modern techniques is not known but is believed to be less.

c. Rhesus isoimmunisation — this complication becomes manifest during a subsequent pregnancy but has been virtually eliminated by the use of prophylactic anti-D immunoglobulin.

d. The risk of ectopic pregnancy is increased if abortion is complicated by sepsis.

e. Higher incidence of babies light for gestation and premature births was reported in one study but has not been confirmed by other large series.

Follow-up

Patients should be seen two to four weeks after termination of pregnancy either in hospital, by the general practitioner, or at a family planning clinic.

1. It is important to confirm that there are no complications such as pelvic infection or continued vaginal bleeding.

2. Pelvic examination should be carried out to ensure that the uterus is empty for very occasionally pregnancy does continue following the procedure.

3. Postoperative counselling may be necessary for those women who are still unhappy or have regrets.

4. Contraceptive advice should have been given before termination and initiated if appropriate before discharge from hospital. The follow-up visit is an excellent opportunity to introduce and reinforce sound contraceptive advice particularly for the less highly motivated patient.

a. If the patient is not already on the pill it can be started immediately.

b. An IUD may be inserted in patients who have had prostaglandin TOP or where the device was not inserted at the time of suction evacuation.

c. An occlusive pessary may be fitted.

d. Arrangements may be made for interval sterilisation if this is desired.

REPEAT ABORTIONS

Repeat abortions account for only 9% of all abortions performed (OPCS 1979). There is no evidence that freely available abortion leads women to rely on abortion rather than use contraception.

Requests for repeat abortion should be considered on exactly the some grounds as for any other case.

It is believed that repeated abortions increase the risk of subsequent obstetrical problems, especially infertility.

RISKS/BENEFITS

The risks associated with abortion depend on many factors — the gestation of the pregnancy, the age, parity, and general health of the woman and the skill of the operator. These risks have to be weighed against the problems for the mother, her family and society of continuing with an unwanted pregnancy. The earlier the pregnancy is terminated, the lower the risk of complications. Termination of pregnancy in the first trimester by vacuum aspiration is a safe procedure (Potts et al 1979) with a mortality of approximately 1 per 100 000. The use of prostaglandins has greatly reduced the risk of complications with second trimester abortions.

REFERENCES

Brewer C 1977 Incidence of post-abortion psychosis: a prospective study. British Medical Journal 1: 476–7.
Office of Population Censuses and Surveys 1974–80, Abortion Statistics Series AB 1–7
Potts M, Diggory P, Peel J 1977 Abortion. Cambridge University Press, Cambridge.
Schulz K F, Grimes D A, Willard Cates Jr 1983 Measures to prevent cervical injury during suction curettage abortion. Lancet i: 1182–1184
Simms M 1973, revised 1977 Nonmedical abortion counselling: Report. Birth Control Trust, London.

Social factors affecting contraceptive use

The social conditions which have led to the need to control births are one of the most obvious factors affecting contraceptive use. Until the beginning of the nineteenth century the growth of the population was rather slow and relatively predictable. The slow growth and population size, here as in other countries, was mainly due to high death rates, particularly among infants and mothers. Birth control as we understand it today was almost nonexistent.

The population 'problem' of our own century has been largely due to the delay and avoidance of mortality without any compensatory decline in the number of births. Improvements in sanitation, housing, hygiene, medical services and knowledge have all contributed. With the resulting massive rise in population, the need for some form of control which was more humanitarian than allowing nature to take its own unbridled course, became more and more obvious.

POPULATION SIZE AND GROWTH

At the present time there are slightly more than 54 million people living in England, Scotland and Wales (Table 13.1) making Britain the 14th most populous country in the world. Due to its relatively small land area, Britain is also one of the most densely populated countries in the world and will continue to be so for some time to come.

The support of so large a population in such a confined space has largely been made possible by the export of manufactured goods and

Table 13.1 Population — numbers and density in the United Kingdom

| | Numbers (thousands) | | | | | | | Population density — number per square kilometre (1981) |
	1901	1931	1961	1971	1979*	1980*	1981	
England	30 509	37 359	43 461	46 018	43 396	46 467	46 221	354
Wales	2019	2593	2644	2731	2775	2778	2790	134
Scotland	4472	4843	5179	5229	5167	5153	5117	65
Great Britain	37 000	44 795	51 284	53 979	54 338	54 398	54 129	235
Northern Ireland	1237	1243	1425	1536	1543	1547	*na*	110†
United Kingdom (Britain)	38 237	46 038	52 709	55 515	55 883	55 945	*na*	229†

Sources: OPCS and General Register Office (Northern Ireland)

*Mid-year estimates. Other figures taken from the Census of Population

†Population density in 1980

na = not available

Note: Differences between totals and the sums of their constituent parts are due to rounding

services. The income from these exports provides the means by which food and raw materials can be bought from abroad to sustain the population. The present population size will not allow a move away from trading for food, even if this is desired.

In Britain today it is not possible to grow enough food to feed the entire population. Even the use of 'marginal' (poorer quality) land would not make up the shortfall. Currently we import 35–45% of all the food and animal feeding stuffs that we need, much of it from countries with growing populations of their own.

It has been estimated that Britain could support itself in food if the population was reduced to between 30 and 35 million, but such a large reduction could not be achieved without the introduction of extreme measures.

This demonstrates the importance of viewing the size or change in a population in terms of the resources that sustain it. Consideration of one without the other has only limited relevance. The most obvious resource is the space the population inhabits, but there are other resources including the fertility of the soil, the climate and the availability of raw materials. The need for a balance between resources and population size assumes the need for self-sufficiency within a particular population, though the British experience suggests that trade between populations may also be seen as a resource. The assumption is that the trade in manufactured goods and services in one direction and food and raw materials in the other will continue. Such an assumption becomes weaker as market forces dictate changes in trading patterns and food-producing countries' need to retain more of their produce to feed their own growing populations.

The moral dilemma

The question of the relationship between personal choice in fertility matters and the needs of the community as a whole, to achieve a desired population size or change is a sensitive one. It is implicitly — and often explicitly — acknowledged in our society that to have babies is a right which, under normal circumstances, cannot be removed. In those societies where this right has been abrogated (e.g. the sterilisation of Jews, gypsies and other 'undesirables' in Germany earlier this century and population 'targets' and coercive sterilisation more recently in India) general condemnation is usual, once the denial of basic rights becomes known.

BIRTH RATES IN BRITAIN

Table 13.2 provides a number of statistics describing the changes that have occurred during the last 20 years in relation to a number of indices that are important to the study of population size. The 754 000 births in 1980 show a rise compared to both 1976 and 1979 but are considerably less than those recorded between 1961 and 1971.

Reasons for change

It is interesting that the recent rise is not due to more women having babies but to women having more babies. If this trend continues, a rise in the rate of population growth is inevitable and the indications that a gradual decline in overall population size may be possible will have been very short-lived.

The reasons for changes in the birth rates are difficult to determine and the availability of contraception is only one of a number of interacting factors. The availability of abortion clearly reduces the birth rate as does the postponement of conception within marriage, the greater time interval between births and the preference for the two-child family.

The influence of social class

Data describing the influence of social class on birth rates have been collected and these are given in Table 13.3. It is of interest to note that in 1980 at least, those in the top two groups (the professional and intermediate classification according to the Registrar General), whilst making up about 25% of the population, produced over 28% of the births. In contrast, the skilled manual workers, who represent 41% of the population produced just under 37% of the total births. Proportionately, it is the most affluent who are producing most offspring with the least affluent coming second, a good way behind.

Table 13.3 also indicates that premarital conceptions leading to live births within marriage are more likely among the semiskilled and unskilled workers. The rate of 27% of all legitimate first live births in this group is more than four times the rate for those in the highest social classification.

These figures do not necessarily reflect the varying incidence of premarital sexual activity across these social groups but they do appear to indicate that when such conception among the working class group

Table 13.2 Some population statistics: United Kingdom

	1961	1966	1971	1976	1979	1980
Live births (thousands)	944.4	979.6	901.6	675.5	734.6	754
Live birth rate (no./1000 population)	17.9	18.0	16.2	12.1	13.1	13.5
Illegitimate live births (no./1000 live births)	57	76	82	90	106	na
Marriages (thousands)	397.1	437.1	459.4	406.0	416.9	414‡
Marriage rate (no./1000 population)	15.0	16.0	16.5	14.5	14.9	14.8‡
Divorce decrees made absolute* (thousands)	25.4	39.1	74.4	126.7	138.7	148.2
Divorce rate* (divorce decrees/1000 married couples)	2.1	3.2	6.0	10.1	11.2	12.0
Deaths (thousands)	631.8	643.8	645.1	680.8	675.5	na
Death rate (no./1000 population)	12.0	11.8	11.6	12.2	12.1	na
Infant mortality rate† (no./1000 live births)	22.1	19.6	17.9	14.5	12.9	na
Neonatal mortality rate† (no./1000 live births)	15.8	13.2	12.0	9.9	na	na
Perinatal mortality rate† (no./1000 live births and stillbirths)	32.7	26.7	22.6	18.0	na	na
Immigration (thousands)	na	219	200	191	195	174
Emigration (thousands)	na	302	240	210	189	229
Balance (thousands)	na	-82	-40	-19	+6	-55

Sources: Population Trends and OPCS

*Statistics relate to England and Wales only

†Definitions: infant mortality — deaths of infants under one year old; neonatal mortality — deaths of infants under four weeks old; perinatal mortality — stillbirths and deaths of infants under one week old.

‡Provisional

na = not available.

Table 13.3 Legitimate births by social class in 1980

Social class	% in population	All legitimate live births		Premarital conceptions**	
		N (thousands)	%+	N (thousands)	%++
I/II* Professional & intermediate	25	164.4	28.4	4.4	6.6
IIIa Skilled nonmanual	11	61.9	10.7	2.6	9.3
IIIb Skilled manual	41	212.3	36.8	15.5	18.7
IV/V Semiskilled & unskilled	20	120.4	21.8	12.8	27.4
Other	3	18.7	3.2	1.7	9.6
All	—	578.9	—	37.0	15.9

Source: OPCS 1982 Birth Statistics
FPA 1982 Fact sheet I.2
*Registrar General's 1980 classification
**Birth within eight months of marriage
+Proportion of total births
++Proportion of all legitimate first live births

occurs, it is more likely to lead to a birth within marriage. These figures indicate that when premarital sexual activity is taking place:

1. The nonmanual workers may be using contraception more effectively.

2. The nonmanual workers may be resorting to abortion more effectively.

3. The nonmanual workers may be disregarding the option of marriage.

Whatever the reason lying behind such findings it seems clear that the working class way of dealing with sexual activity that leads to pregnancy is to get married, whereas the middle class way is to use contraception effectively and if pregnancy occurs, to seek an abortion. It is possible the working class couple sees real or imagined obstacles to the use of contraceptive or abortion services which are not so obvious to the middle class couple.

ABORTION AND ILLEGITIMATE BIRTHS

Approximately 40% of all pregnancies conceived outside marriage end in legal abortion compared to 8% within marriage. Table 13.4 provides an indication of the number of abortions and illegitimate births for England and Wales in 1979 and in 1980 and also for Scotland during 1979. The ratio of illegitimate births to live births conceived within marriage in England and Wales (1:9) and in Scotland (1:10) and the ratio of abortions to live births (1:5 and 1:9 respectively) were more favourable to Scotland in 1979. The figures for 1980 in England and

Table 13.4 Abortion and illegitimate births

	England and Wales 1979	1980	Scotland 1979
Illegitimate live births	69 460	77 400	6960
(women under 20 years	24 000	26 100	2450)
Abortions	120 600	126 000*	7700
(women under 20 years	32 700	31 200**	2100)

Source: OPCS Abortion monitors
        OPCS 1982 Birth statistics, series FMI No 7
*Comprising 82 000 outside marriage and 44 200 inside marriage
**Comprising 30 500 outside marriage and 700 inside marriage

Wales suggest a worsening of these ratios for 1 in 7 births were illegitimate and there was approximately 1 abortion for every 4 live births. It is likely that the figures for illegitimate births and abortions in Scotland will also indicate an upward trend but it is unlikely these will be at the same level as those for England and Wales.

THE USE OF CONTRACEPTION

There is considerable difficulty in obtaining data describing the overall use being made of different contraceptive methods in Britain. No national figures are compiled and small-size surveys are often used to estimate this information. In addition, estimates can be based on the manufacturers' sales data but these apply only to some contraceptives and it is uncertain if the sales reflect actual use. In 1982 the Family Planning Association published estimates of contraceptive use and these are given in Table 13.5.

Table 13.5 Estimated contraceptive usage in the UK (Women aged 15–44)

	1978	1980
Oral contraceptives	3 100 000	3 018 000
Male and female sterilisation	600 000	—
Condoms	2 700 000	2 801 000
IUDs	600 000	498 000
Diaphragms	300 000	192 000
Coitus interruptus	700 000	—
Rhythm methods	200 000	—
Spermicides used alone	100 000	36 000
Injectables (Depo-Provera)*	33 000	—
Total	8 333 000	

Source: FPA 1982 Fact Sheet C.2(31), January
*This is a whole year figure 1977. The half-yearly figure for injectables for 1978 was 19 000

Unplanned pregnancy, marriage and divorce

It is not unreasonable to assume that almost all of the pregnancies resulting in abortion and most of the pregnancies ending in illegitimate births, were unplanned. The numbers of full-term births taking place within eight months of marriage also suggests a high proportion of unplanned pregnancies but this is harder to prove. The incidence of cohabitation before marriage is increasing year by year but this does not necessarily indicate a disillusionment with marriage, for marriage often takes place once a pregnancy occurs. It is the timing of marriage in relation to sexual activity and reproduction that is changing. Although divorce has increased to about one in every three first marriages, the incidence of remarriage is also rising. It is not so much the concept of marriage that is failing but that the expectations surrounding the person to whom one is married at any particular time are not being met. The effects of changes in marriage patterns and divorce legislation will have a profound effect on fertility levels and therefore the use made of contraceptive products.

Method acceptability

Although there are over 8 million contraceptive users, the Family Planning Association frequently makes reference to the 200 000 unplanned pregnancies that still take place each year in this country. This large number continues despite the introduction of modern methods of contraception which can be virtually certain of preventing pregnancy (if they are correctly used).

Here is the basis for the faulty reasoning which lies behind many family planning programmes and services. It is incorrectly assumed that all that is required is the provision of a contraceptive which is effective and which does not create unwanted side effects. If effectiveness and safety are maximised then the 'perfect' contraceptive has arrived. This reasoning ignores the most important factor of all, namely the acceptability of the method to the consumer, customer or patient.

It is frustrating to those who produce what they consider to be an effective and safe method of contraception to find that those for whom the method is intended will not use it, or give it up after only a short time. Perhaps this is why those who seem most difficult to help are sometimes labelled as being 'fertile and feckless', or without motivation. Of course, the concept of acceptability includes ideas about effectiveness and safety but it is much more broadly based than these two factors alone and varies significantly between different social groups. Without an understanding of contraceptive acceptability an effective service is unlikely to be achieved.

Perception of the method

If we assume that every patient's perception of the condom, cap, pill, IUD and so on is much the same then we are unlikely to be able to appreciate the real reasons why individuals may not use them or may hurriedly give up their use.

For example, the methods of contraception which are perceived as suitable by an older parous woman may appear entirely inappropriate to a younger nulliparous woman attending the contraceptive service for the first time. Also the influence of social factors is felt most strongly in relation to the perception of appropriate behaviour and the feelings surrounding the use of specific contraceptive methods.

Characteristics of the method

One way of being more precise about what determines the acceptability of a contraceptive method is to examine in some detail its characteristics, looking not only at its effectiveness and safety but also at the behavioural components related to its use. Behavioural questions include the following:

1. Which partner used the method?
2. How often is it used?
3. Is it coitally dependent or independent?
4. How is it delivered into the body?
5. What part of the body is affected?
6. What ancillary equipment is necessary?
7. For how long is it effective?

This list could be much longer. Clearly the question of male versus female method depends on the perception of the male and female role in preventing conception. To a very large extent this perception is socially determined.

The need for daily pill-taking by oral contraceptive users, even when sexual intercourse is not taking place, is in marked contrast to the far less frequent action needed to be taken by a patient with an IUD. Remembering to take the pill in circumstances when one's feelings and thoughts are far removed from sexual activity may not prove acceptable, especially when a fear of unpleasant side effects is also present. In such circumstances it is not the pill itself that is making the method less acceptable, but the daily reminder of one's own uncertainty.

The dual problem of using a method that may not be needed immediately and using a coitally-dependent method, which inevitably interrupts the sex act itself in some way, is virtually impossible to resolve without a thorough understanding of the relationship between the couple concerned. Similarly, the route of administration is a very important factor influencing the acceptability of a method. The relative acceptance of methods administered through the mouth, through the skin or directly into or through the vagina have varying degrees of acceptability depending on the cultural and personal environment to which the person concerned is inevitably linked. What is politely described as 'digital exploration of the vagina' is not acceptable in some sociocultural groups or to certain individuals.

Research can be undertaken without specific reference to contraceptive technology. Knowing how women feel about examining themselves vaginally may appear to be very narrowly focused but it says much about the acceptability of such behaviour in relation to the fitting and removal of caps and diaphragms, use of foams, jellies and creams and even the feeling for the IUD tails to ensure correct placement of the device. Seen in this light, the apparently unrelated assessment of personal vaginal exploration becomes a very important feature when gauging the acceptability of various existing and potential methods of contraception. By ascertaining the meaning of such behaviour to those concerned it may be possible to discover why some methods are more, or less, acceptable than others (Zeidenstein 1980, Bruce & Shearer 1979).

Nonmedical methods

Another look at Table 13.5 shows that if the numbers using the condom, coitus interruptus, rhythm and spermicides are combined, the sum of 3 700 000 users represents 44% of the total. This very large group is using methods that have two major features in common: they have virtually no side effects and they do not necessarily require attendance at a family planning clinic or doctor's surgery. A third characteristic is that, with the exception of the condom, they have relatively low effective rates in preventing pregnancy. For millions of couples the effectiveness of their chosen contraceptive method appears less important than its more general acceptability and safety.

REFERENCES

Bruce J, Schearers B 1979 Contraceptives and common sense: conventional methods reconsidered. Population Council, New York

Zeidenstein G 1980 The user perspective: an evolutionary step in contraceptive service programes. Studies in Family Planning, vol 11, no 1, p 24–29.

Medicolegal aspects of family planning

All civilised countries are under the rule of law. In the United Kingdom the law is in general governed by:

1. Common law, established over centuries.

2. Statute law, embodied in Acts of Parliament and modified by regulations.

3. Case law, the courts relying on judgements given in previous similar cases.

Here there is space to deal only with the law in the United Kingdom and it must be pointed out that the law in Scotland differs in many respects from that in England and Wales. There are also differences in Northern Ireland, where, for example, the Abortion Act (1967) does not apply.

THE LAW RELATING TO CONTRACEPTION AND STERILISATION

Methods of contraception

There are no legal restraints on the provision of mechanical means of contraception such as caps, condoms and chemicals for vaginal use.

These may be bought freely over the counter in chemists' shops and elsewhere, or be provided in family planning clinics.

There are restraints on the provision of hormonal contraceptives which must be prescribed by a registered medical practitioner.

The position regarding intrauterine devices is unclear. The sterilised pack containing the Lippes Loop states 'to be sold only on the order of a doctor' but there would appear to be no legal restraint on the provision or fitting of an intrauterine contraceptive.

Postcoital contraception may be used as an emergency measure where unprotected intercourse has taken place at or near the time of ovulation. It may also be considered in cases of an accident with contraception, such as a sheath bursting, or in cases of rape (Ch. 11).

The question as to whether postcoital contraception is illegal was put to the Director of Public Prosecutions. The Attorney General ruled that it is not illegal, so that any doctor who administers postcoital contraception is not in danger of prosecution. As pregnancy cannot be said to exist before implantation, postcoital contraception may be held to constitute interception rather than abortion.

Sterilisation

It is generally held that an operation for sterilisation is not illegal provided it is performed for a valid reason and with due consent. But Mr Justice Woolf ruled (see below) that a girl under the age of 16 probably cannot give consent to an operation for sterilisation.

Consent

Consent itself may be written or implied. When a patient attends a clinic or visits a doctor it is implied that consent is given to any reasonable procedure. Written consent should be obtained for any procedure involving an anaesthetic or where a particular risk is involved.

Patients aged 16 and over

Section 8 of the Family Law Reform Act (1969) provides that the consent to medical or surgical treatment of a minor who has attained the age of 16 years shall be effective consent and that in such cases it is not necessary to obtain consent from parent or guardian.

In Scotland age of consent is covered by common law. Any minor (over the age of 14 for boys and 12 for girls) has substantial legal capacity.

Patients under the age of 16

The possibility has been raised that a doctor who gives contraceptive advice to a girl under 16 may be aiding and abetting the offence of unlawful sexual intercourse. Both the General Medical Council and the British Medical Association have advised that the doctor should endeavour to persuade the girl to involve her parents or guardian from an early stage in the consultation. If the girl refuses the doctor should prescribe, if this is in her best interests, and respect her confidentiality. A similar view was expressed in an NHS notice (HN80 46) issued in December 1980.

This view was challenged in the High Court by a mother who sought declarations that the NHS circular was giving unlawful advice. Mr Justice Woolf ruled that the fact that a child is under the age of 16 does not mean automatically that she cannot give consent to any treatment. He did add that he thought it unlikely that any child under 16 could consent to an operation for sterilisation. He therefore rejected Mrs Gillick's case. However, at the time of writing, there may be an appeal against the judgement (Brahams 1983).

The mentally subnormal

A request may be received to carry out sterilisation on a mentally subnormal individual. This is a difficult situation as no operation can legally be performed without informed consent. In other words the patient must understand the nature and possible consequences of any surgical procedure.

If a patient suffers from mental subnormality of such severity that informed consent cannot be given, and this will apply even to those under the age of 16, such an operation might be held to constitute an assault and the patient might be entitled to damages. In such cases the doctor would be wise to consult his defence society before proceeding further.

Consent of the partner

The question of the consent of the partner will only arise if:
1. The couple are legally married.

2. The couple are living together.

3. The husband is supporting his wife and children.

In such cases it is usual and probably wise to seek the consent of the partner if an operation for male or female sterilisation (including all operations such as hysterectomy, which may result in sterility) is to be undertaken. The legal position is not absolutely clear but it is doubtful if a husband would have any right of action if a consenting wife were sterilised without his consent. A similar position will arise with vasectomy.

It may be assumed that each partner in a marriage has a reasonable right to have children by the other. Hence many clinics used to ask for the agreement of the husband before inserting an intrauterine device into a married woman. However, a recent decision by a court which held that a woman did not need her husband's consent before her pregnancy was terminated has raised doubts as to the need for this (see below).

A husband's permission is not required before the pill is prescribed for his wife or an occlusive pessary fitted.

Consent to treatment

All consent to treatment must be *informed* consent. Thus Mr Justice Russell recently awarded damages to a woman who complained that she had been given a contraceptive injection (Depo-Provera) without her consent. She sued for damages for trespass and assault. The nature and possible side-effects of such an injection should always be explained to the woman who should be given the opportunity to refuse it. A careful note should be made that such an explanation has been given.

THE LAW RELATING TO ABORTION

The law relating to abortion in England and Wales is laid down in the Offences Against the Person Act of 1861, Sections 58 and 59. This act was consolidating legislation embodying previous acts and dealing with a number of offences against the person. It did not apply to Scotland. Under the 1861 Act it was an offence punishable by up to life imprisonment to attempt to terminate a pregnancy by artificial means, whether by the woman herself or by another person and whether the woman was or was not pregnant.

The Abortion Act (1967)

The present legal position in Great Britain (but not in Northern Ireland) is governed by the Abortion Act (1967) which came into force in 1968. The 1861 Act is still in force but the 1967 Act states that:

1. 'A person shall not be guilty of an offence under the Law relating to abortion when a pregnancy is terminated by a registered medical practitioner if two registered medical practitioners are of the opinion, formed in good faith —

   a. that the continuance of the pregnancy would involve risk to the life of the pregnant woman, or of injury to the physical or mental health of the pregnant woman or any existing children of her family, greater than if the pregnancy were terminated: or

   b. that there is a substantial risk that if the child were born it would suffer from such physical or mental abnormalities as to be seriously handicapped'.

2. In reaching the decision as to whether termination of pregnancy is to be performed, 'account may be taken of the pregnant woman's actual or reasonably forseeable environment'.

3. Treatment for termination of pregnancy may only be undertaken in a National Health Service (NHS) hospital or in a place approved by the appropriate Secretary of State. In practice such clinics are inspected regularly on behalf of the Department of Health of each country.

4. The certificate given by two registered medical practitioners must not be destroyed for three years.

5. Every procedure for the termination of pregnancy must be notified within seven days to the Chief Medical Officer (CMO) at the Department of Health or to the relevant CMOs in Scotland and Wales. This is a confidential notification on a 'doctor to doctor' basis (administrative staff in these offices are bound by confidentiality) and may not be revealed to anyone. The purpose of notification is to keep records relating to the different regions of the country and for research purposes. Regulations under the Act permit the CMO to divulge the information in certain circumstances, e.g. to determine whether an offence has been committed.

6. The emergency clause permits a registered medical practitioner to terminate a pregnancy in an emergency situation without the opinion of another practitioner. Such cases must be notified.

7. The conscience clause states that no person shall be under any duty to participate in any treatment under the Act to which he has a conscientious objection, except that in an emergency situation anyone

has a duty to give treatment to save life or prevent serious injury to health.

Menstrual extraction

Menstrual extraction, or menstrual evacuation, is a method used in cases where there may be a possibility of unwanted or unplanned pregnancy. The uterus is evacuated using a small curette and a suction aspirator at the time menstruation is due or within a few days. Legal opinion has been expressed that this would constitute an attempt at abortion under the 1861 Act and that, therefore, the conditions laid down in the Abortion Act 1967 should be complied with. The situation would be different in Scotland where the 1861 Act does not apply.

Consent to termination of pregnancy

The general rules of consent to treatment apply to procedures for the termination of pregnancy.

1. If the woman is aged 16 or over and is either unmarried or not living with her husband and he is not supporting her, she can give valid consent. She is entitled to an explanation of the nature and effects of any procedure which is contemplated and this should be given by a doctor or a senior nurse if no doctor is available.

2. It has been usual to get the consent of the husband if the couple are legally married and living together but the legal necessity for this has been questioned by a decision of the court which held that a woman did not need her husband's consent before her pregnancy was terminated. However, in the case in question the couple were not living together and had commenced divorce proceedings. Until the law is clarified it is probably wiser to obtain the consent of the partner before termination of pregnancy is undertaken. If, however, this consent is unreasonably refused and there are medical grounds for termination of pregnancy the operation could be carried out without the consent of the partner.

3. In the case of girls under the age of 16 consent of the parent or legal guardian should be obtained before any operation is undertaken. In the case of a minor in the care of the local authority, the staff of the social services department have the right to give or withhold consent in respect of any child in their care. On the other hand, a girl under 16 should not be pressurised by parents or legal guardians to agree to a termination if she wishes to keep the child.

The guardianship function of the local authority in Scotland is exercised through the social work department under the Social Work (Scotland) Act 1968.

4. The mentally subnormal. The same criteria apply as for the provision of contraception and for sterilisation (see above).

5. It should never be made a condition for termination of pregnancy that the woman consent to sterilisation at the same time. Indeed, there are good reasons for deferring sterilisation until the woman has recovered from the effects of termination of pregnancy so that she can come to a considered decision.

The Abortion Act has been justified if only in the dramatic reduction that has resulted in deaths from abortion and in the virtual disappearance of prosecutions for criminal abortion.

The legal requirements of the Act must be complied with but there remains the need for careful and sympathetic counselling, not only of the women with unwanted pregnancies but of their partners and families who are often involved in distressing situations.

THE ETHICS RELATING TO COMMUNICATION BETWEEN PROFESSIONAL COLLEAGUES AND TO MEDICAL RECORDS

In the United Kingdom, every individual man, woman and child is entitled, under the provisions of the NHS, to be registered with a family doctor or general medical practitioner. Referral to a consultant or specialist, whether in hospital or in private practice, should take place in most cases through the family doctor.

A different situation arises in the case of 'open clinics', such as family planning, health screening or immunisation clinics. Patients may refer themselves to these clinics but it is desirable that the family doctor be kept informed of the results of investigations or any treatment given. In the case of family planning clinics it is clearly important for him to know if his patient is taking the contraceptive pill. An exception exists in the case of clinics for the investigation and treatment of venereal disease (genito-urinary medicine). Here patients are treated with strict confidentiality.

In most cases doctors do not work in isolation and there must be, of necessity, many people who will have access to confidential information concerning patients — medical records staff, medical secretaries, receptionists, nurses, midwives and pharmacists to name only a few. Since the days of Hippocrates it has been held that everything which passes between a doctor and a patient is confidential and may not be revealed to anyone except in certain well-defined circumstances.

The same duty of confidentiality rests upon all who have access to information about patients.

Referral

A general practitioner may refer a patient for advice to a consultant or a specialist. A request for a second opinion or for referral to a specialist should not be unreasonably refused.

The letter of referral may be held to be the subject of qualified privilege and may not, for example, be used as material for any action for libel. If there is any doubt as to whether a referral letter might be libellous it should be sent direct to the consultant and not given to the patient.

The consultant in his turn should report the results of his examination, and of any investigation or treatment recommended or carried out, to the general practitioner. Proper communication between professionals advising patients is essential.

Case records

1. In general practice the records of NHS patients are the property of the appropriate Secretary of State but the doctor holds them in law and has the duty of preserving their confidentiality. He should pass the records on if the patient transfers to another general practitioner.

2. Records of private patients are the property of the individual doctor.

3. In NHS hospitals and family planning clinics the patient's records are the property of the Department of Health (or the relevant department in Scotland, Wales and Northern Ireland).

4. Records of social service departments are the property of the local authority. As such, they may be inspected by local councillors.

Confidential information

Doctors are often invited to reveal confidential information concerning patients.

1. The request may come from relatives or friends who are reasonably concerned about the patient's condition. It is generally unwise to divulge confidential information without the patient's

consent but this consent may be verbal and the doctor must use discretion in deciding how much to reveal.

2. Requests from journalists or the media should always be refused.

3. The doctor may be asked to give information concerning a patient to the police when it is suspected that an offence has been committed. In such cases it is generally wise to say simply that it is not possible to reveal information obtained in confidence, except by order of a court.

4. Information should not be given to an insurance company without the written consent of the patient and then it is best given to the company's medical adviser on a 'doctor-to-doctor' basis.

5. A doctor may also be asked to give confidential information to lawyers in relation to matrimonial cases or to proposed actions for negligence. A doctor was recently found guilty of serious professional misconduct by the General Medical Council because he revealed confidential information about a woman patient to her husband's lawyer in a divorce case without her consent. In the case of an action for negligence it has been agreed that medical records concerning the case may be sent to a medical adviser nominated by the other side.

6. If a doctor is summoned to a court to give evidence he will be compelled to reveal confidential information under oath and to produce his records. Before doing so he must make it clear that the information requested has been obtained in confidence.

RESPONSIBILITY OF THE DOCTOR

It has already been stated that the doctor owes to any patient who consults him the duty of confidentiality. The other responsibility of the doctor towards his patient will depend on the circumstances in which the patient consults him.

In the specialist hospitals and clinics of the NHS and in family planning clinics, the doctors employed there probably have a duty to see any patient who is referred to them or, in the case of 'open door' clinics, any patient who seeks their advice. Accident and emergency clinics also generally see any patient who seeks treatment though the patient may be advised to consult his/her general practitioner when the emergency or the immediate situation has been dealt with.

A general practitioner

1. Is required to provide general medical services which include the prescription of drugs, certification and referral for specialist advice where this may be indicated.

2. Is not obliged to accept any patient on his list unless there is no other practitioner within reasonable reach of the patient's residence and may ask the patient to transfer to another practitioner without giving any reason.

3. May decline to provide family planning services for his own patients who, while still remaining on his list, may then seek advice from another source including another general practitioner (Ch. 1).

4. Is expected at all times to exercise due care and skill in the light of his own training and experience. When a patient suffers damage or loss as the result of a negligent act by a doctor she may have a claim in law against the doctor and be entitled to receive damages.

5. Is expected to provide the patient with a reasonable explanation of all treatment given or procedures recommended to which the patient has the right to refuse.

Doctors engaged in family planning are in the same situation of responsibility as all other doctors

1. They are required to exercise due care and skill in their dealings with patients, to carry out adequate examination and such investigations as may be available.

2. They should give an adequate explanation of the case to the patient, recommend treatment and ensure that the patient understands the nature of the treatment, how it should be taken and its possible side effects.

3. They have a duty to keep careful records and to ensure the confidentiality of all information given.

4. Where there are no facilities or equipment for complete investigation and treatment of a problem it may be wise to refer the patient for a specialist opinion. The patient's general practitioner should always be informed of the results of investigations and of the treatment recommended unless the patient specifically refuses to have such information passed on. This refusal should be recorded in the patient's notes.

RESPONSIBILITY OF THE PATIENT

1. The patient gives implied consent to any necessary physical examinations and investigations such as blood tests which may be necessary for the management of her case.

2. If sterilisation is recommended, written consent from her and probably from her husband will be required (see above). Any advice given should be fully accepted by the patient. In discussing sterilisation the remote possibility of failure of all methods must be made clear to both partners.

3. She is entitled not to accept the advice given but if she does not carry out the recommendations of her advisers they cannot be held responsible.

4. The patient who discharges herself from hospital care against advice should be asked to sign a statement to this effect or, if she refuses, a note should be made in her case records.

Most patients are anxious and willing to accept advice they have sought. In all matters concerned with family planning the importance of counselling cannot be overemphasised.

THE CLINIC TEAM

In the modern medical scene, doctors do not work in isolation. The family planning team consists of many individuals — doctors, nurses, receptionists, secretaries, telephonists and others.

1. Doctors working in family planning clinics will be expected to have had special postgraduate training — to be a Fellow or Member of the Royal College of Obstetricians and Gynaecologists, or to hold the certificate of the Joint Committee on Contraception.

2. At present general practitioners are not required to undertake any such special training before providing family planning services in their practice.

3. Nurses may take special training in the form of Course 900 (Ch. 1).

Delegation of duties

A great deal of work in clinics may be delegated to nurses provided they are properly trained for the task.

1. Discussion of the methods of contraception available and the teaching of their use.

2. Fitting caps.

3. Taking cervical smears.

4. Issuing the contraceptive pill which has been prescribed by the doctor. At present the pill can only be legally prescribed by a medical practitioner who takes the responsibility for any medication he prescribes. If he does not exercise due care and the patient comes to

any harm he will be held liable. In many clinics nurses dispense repeat supplies of the pill under the signature of the doctor and the legality of this has not so far been questioned. It is not legal for a nurse to issue pills to a patient unless the doctor has already signed the prescription.

It has been proposed that properly trained nurses should be permitted to prescribe or at least supply the pill in clinics without medical prescription. It has also been suggested that the pharmacist should be allowed to dispense the pill without a doctor's prescription as applies in some countries. However, as the law stands in the United Kingdom, the pill remains on prescription by medical practitioners only.

5. Removal of intrauterine devices.

6. Under special circumstances, and after proper training, in some clinics nurses fit intrauterine devices.

The nurse will be legally liable for anything she does that is properly the work of a nurse without medical supervision. Many nurses take out independent insurance to cover them in case of legal liability for any mishap that may come to the patient.

THE ROLE OF THE DEFENCE SOCIETIES

Every medical practitioner faces a medicolegal risk in the course of his professional life. During the 1880s the increase in attacks on medical practitioners and threats of prosecution, often on frivolous pretexts, led to the formation in 1885 of the Medical Defence Union. This was followed by the London and Counties Medical Protection Society (now the Medical Protection Society) and the Medical and Dental Defence Union of Scotland.

All three organisations have similar objects:

1. To protect the character and interests of doctors and dentists.

2. To promote honourable practice and to suppress or prosecute unqualified practitioners.

3. To assist and defend members involved in litigation.

4. To advise members on any matter affecting their professional character or interest and to indemnify those involved in legal claims.

It is sometimes assumed that the defence societies are insurance companies which insure doctors against claims for negligence. This is not strictly so since, although the societies will readily give legal advice and will pay legal charges and damages in negligence actions, they are not obliged to do the latter, each case being considered on its merits.

A large part of their work consists of advising members on matters of medicolegal concern. They do not normally defend doctors (or dentists) involved in criminal actions unless a professional principle is at stake, but they will usually defend doctors accused of serious professional misconduct before the General Medical Council.

All doctors should, and in most cases must, as a condition of their employment belong to a defence society. This applies even to a doctor who does only one or two sessions a week or occasional locums in family planning since he is still at risk of having to face a legal action for negligence and an expensive claim.

If a serious complaint, a threat of legal proceedings or a claim for compensation is received, the doctor:

1. Should inform his defence society at once.

2. Should not attend any conference with lawyers or administrators without consulting the society.

3. Should not consult private solicitors without consulting his society because if he does so, further help may be reasonably denied him.

It is important that all medical work should be done with scrupulous care. Due consent with adequate explanation should be obtained before any operative procedure is undertaken. Patients should be adequately examined and, above all, careful records kept to include any correspondence relating to the patient. If a patient will not accept the medical advice offered this should be recorded. Case notes should not be altered retrospectively or, if they are, a note should be made to that effect. It must be remembered that in the case of a threat of legal action the records may be released to an expert appointed by the other party or produced in court.

In the context of modern medical practice, medical defence is becoming of increasing importance. There is much litigation, some of it very expensive as damages and legal costs escalate. Many cases now have to be settled out of court and when there has been obvious negligence on the part of a doctor the defence societies will prefer to compensate the patient rather than fight, and possibly lose, a potentially expensive legal action with all the attendant unpleasant publicity for the doctor concerned.

DOCTORS AND THE COURTS

A medical practitioner, like any other citizen, may be required to give evidence in court if he has witnessed an accident or a suspected criminal offence.

Expert witness

The doctor is in a special position as an expert witness. His training and knowledge are used to assist the court in determining the medical facts of a case.

In cases where an unnatural death occurs a doctor may be summoned to give evidence in a coroner's court in England, or in Scotland to report to the Procurator Fiscal and give evidence at a fatal accident enquiry. In such a case he should take with him the relevant case records and should ask for permission to consult them in the course of his evidence. If there is any likelihood that an accusation may be made against him or a claim for negligence result he should consult his defence society before attending.

In actions for negligence held in the High Court (or the Court of Session or Sheriff Court in Scotland) it is usual for there to be a conference with counsel at which the details of the case will be studied. The doctor, as an expert witness, should have a copy of all relevant case notes and should produce books or journals likely to help the Court.

In giving evidence it is important:

1. To address the judge, or, in a lower court, the magistrate (in Scotland the Sheriff).

2. To speak slowly and distinctly as the judge will be taking notes.

3. To keep answers short.

4. Not to make a speech to the court.

If he is an expert witness on behalf of a plaintiff (or pursuer in Scotland) the doctor will first be questioned by the plaintiff's counsel, then cross-examined by counsel for the defence. The plaintiff's counsel may re-examine if there are points to be clarified.

NULLITY

There are various circumstances in which a marriage may be annulled. One which is important medically is where there has been wilful refusal or incapacity on the part of either party to consummate the marriage. In the circumstances of a family planning clinic, a doctor may be asked to give a 'certificate of virginity' in a case of nonconsummation of marriage. It is, in general, unwise to do this as a proper legal procedure exists for the handling of such cases. The party concerned should be advised to consult a lawyer. Legal aid may be available to necessitous persons. In practice there are throughout the country appointed inspectors in nullity and the lawyer will refer the parties to one or more of these. Procedure in nullity cases has been much simplified in recent years; if the case is undefended, which most are, a statement from the inspector will generally be accepted by the Court and it will not be necessary for the inspector to attend to give evidence.

ACKNOWLEDGEMENT

I should like to thank Dr Brooke Barnett, Secretary of the Medical Defence Union, who read the manuscript of this chapter and made many helpful comments.

REFERENCES

Abortion Act 1967 HMSO, London, 1967.
Brahams D 1983 Confidentiality and under-age girls who seek contraceptive advice. Lancet ii: 177.
Brahams D 1983 Under-age girls and contraception: the parent's right to be informed. Lancet ii: 350–352.
Social Work (Scotland) Act 1968 HMSO, London, 1969.

15

Patricia Last

Health screening at the family planning consultation

Screening

Screening for pelvic disease
Pelvic cancer: carcinoma of the cervix
  Risk factors
  Diagnosis
  Prevention of invasive cervical cancer
  Taking a cervical smear
  Cervical intraepithelial neoplasia,
    carcinoma in situ
  Management of women with abnormal
    smears
  Frequency of cervical smears
Cancer of the ovary and corpus uteri
Other pelvic conditions
  Cervicitis
  Cervical erosions
  Cervical polyps
  Ovarian cysts
  Uterine fibroids
  Retroversion of the uterus
  Sexually transmitted diseases

Screening for breast cancer
Screening techniques
  Breast self-examination
  Clinical examination
  Mammography
  Management of the patient with a breast
    lump

Other screening
  Weight
  Blood pressure
  Urine testing: haemoglobin estimation
  Rubella screening

Who should screen?

Health promotion
  Smoking
  Exercise

Appendix

Apart from routine medical examinations at school, most women do not have any physical examination until they attend for family planning advice in their teens or early twenties. The family planning consultation therefore offers an excellent opportunity for screening and for promoting health.

The supervision necessary for women using various contraceptive techniques has been discussed in the relevant chapters. Screening will be considered as a separate entity although in many cases the examinations will overlap. The screening procedures recommended in this chapter represent a counsel of perfection. No-one should be refused contraception just because she does not wish to avail herself of the screening facilities offered. The approach of the clinic personnel should not be so authoritarian that an overweight smoker feels too guilty to attend. Even the vaginal examination and smear test recommended at the first visit can often be omitted if the patient feels unable to accept them.

SCREENING

Health screening is the examination of sympton-free persons in an attempt to discover early disease or the predictors of disease. The examination of those with symptoms is disease diagnosis and is quite separate.

The theory of screening is based on the belief that disease caught early in its progress or even before it is clinically manifest will respond more readily to treatment, with lower morbidity and mortality rates than disease left to become clinically well established and symptomatic.

The *outcome* of many diseases depends on four factors:

1. The aggression of the disease.
2. The resistance of the host.
3. The state of the disease at the time of diagnosis.
4. The treatment available to, offered to, or accepted by the patient.

Screening is concerned with the third parameter. Not all diseases will respond favourably to early detection and certain rules are used as guidelines for a screening programme.

The disease to be screened for must be:

1. Serious enough to justify the search.
2. Common enough to give a reasonable pick-up rate.
3. Developing slowly with a fairly long presymptomatic phase.
4. Amenable to treatment which must be available and acceptable to the patient. Furthermore, there must be an advantage in treating the disease at a stage before the patient would otherwise present.

The screening test must be:

1. Simple and easily applied.
2. Positive in the majority of persons with the disease or predictors of the disease. This demands a high sensitivity, i.e. a test producing few false negatives.
3. Negative in the majority of persons without the disease. This demands a test of high specificity, i.e. one producing few false positives.
4. Acceptable and not harmful to the patient.
5. Economically possible with the current resources of the health care service in the country concerned.
6. Capable of giving reproducible results independent of observer variation.

The *prevalence* of a disease is the finding of an abnormaility in a well-defined population examined for the first time.

The *incidence* of a disease is the finding of new cases in the same screened population when examined at specific intervals.

SCREENING FOR PELVIC DISEASE

Pelvic cancer: carcinoma of the cervix

Cancer in women under the age of 45 is uncommon. The exception to this statement is carcinoma in situ of the cervix, which has a peak incidence in the 25–34 age group. The incidence of genital tract cancer rises dramatically in the age group 45–54, but most women will have

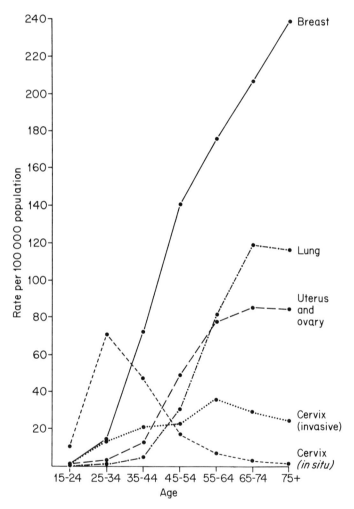

Fig. 15.1. Estimates of cancer registrations 1980 for women for England and Wales (OPCS Monitor MBI 83/1).

left the supervision of the family planning clinic by that time (Fig. 15.1).

The cervix is the second commonest site for invasive cancer in the female genital tract (Table 15.1). Some 2000 women die each year from this disease — just under 4% of all cancer deaths in women. The death rate from cervical cancer is falling. The registration rate for carcinoma of the cervix is also falling, though by a very small margin, and fewer cases are recorded each year (Tables 15.2 and 15.3).

Table 15.1 Estimated cancer registration rates per 100 000 women: England & Wales 1980 (OPCS Monitor MB1 83/1)

	Lung	Breast	Cervix (inv.)	Cervix (in situ)	Corpus uteri	Ovary
15–24	0.1	0.7	1.3	10.9	0.1	1.0
25–34	0.6	14.8	14.6	71.0	0.7	3.0
35–44	5.5	71.9	21.7	46.8	3.1	9.8
45–54	30.6	141.5	22.5	17.4	20.4	28.2
All ages	34.8	85.9	16.2	19.7	14.0	17.8

Table 15.2 Cancer death rates per 100 000 women age standardised: England and Wales. Cancer Research Campaign 1982. Trends in Cancer Survival in Great Britain

	Lung	Breast	Cervix (inv)	Corpus uteri	Ovary
1961–1965	13.2	34.6	9.4	4.2	11.4
1966–1970	16.7	36.1	8.8	3.8	12.1
1971–1975	20.2	38.4	7.8	3.6	12.2
1976–1978	23.0	39.4	7.6	3.5	12.5

Table 15.3 Cancer registration rates per 100 000 women age standardised: England and Wales. Cancer Research Campaign 1982. Trends in Cancer Survival in Great Britain

	Lung	Breast	Cervix (inv)	In-situ	Corpus uteri	Ovary
1962–1965	12.7	58.9	—	—	—	12.9
1966–1970	15.9	63.6	16.8	11.7	11.1	13.6
1971–1974	20.1	71.6	15.6	11.5	11.3	14.3

Risk Factors

The most important factor is sexual activity. Squamous cell carcinoma, responsible for 95% of all cervical cancers, is unknown in virgins.

The incidence is increased in women who have many sexual partners or who associate knowingly or unknowingly with men who have had many partners. Those who start sexual activity at a young age, when the cervical epithelium may be more vulnerable to carcinogens, are at high risk. Various carcinogens are being currently investigated — the herpes simplex viruses, the human papilloma (wart) viruses, and particular components of human semen.

Diagnosis

The diagnosis of cervical cancer can usually be made on the appearance of the cervix on speculum examination. It is a rare condition and even a family planning doctor will see less than five cases in the course of a professional lifetime. Taking a cervical smear may be helpful in diagnosis, but in 10–15% of histologically proven cervical cancers, smears fail to show any neoplastic cells. Any patient with postcoital or irregular intermenstrual bleeding should be referred for further investigation if the condition is not cured by simple remedies, e.g. removing an IUD, stopping the pill or treating an infection.

In 1963, the World Health Organisation stated that invasive carcinoma of the cervix is a preventable disease. If preventable, why not prevented?

Prevention of invasive cervical cancer

The great majority of invasive cervical cancers are preceded by a neoplastic noninvasive process — cervical intraepithelial neoplasia (CIN) — which cannot be diagnosed by the naked eye but which may be detected by exfoliative cytology. Since CIN is virtually 100% curable, cervical screening by smear test must rank as one of the most important screening procedures for women during the family planning consultation. The registration rates for cervical cancer are falling, particularly in countries with extensive screening programmes. Data from the city of Aberdeen, where a very active screening programme is pursued, provide further support to the premise that the incidence of and deaths from cervical cancer can be considerably reduced by intensive screening (Fig. 15.2, Table 15.4).

Taking a cervical smear

In theory this is an easy procedure (Macgregor 1981), but in practice it seems to present difficulty. Approximately 10% of smears are

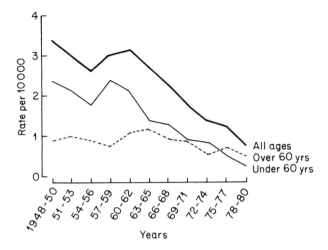

Fig. 15.2. Incidence of invasive squamous carcinoma of uterine cervix, city of Aberdeen 3-year average per 10 000 population 20+ years (1948–1980) (Macgregor 1981).

unsatisfactory and have to be repeated. The commonest error is failing to take an adequate sample from the squamo-columnar junction, the area where neoplastic change occurs. In nulliparae, and in women past the menopause, the squamo-columnar junction may be well within the cervical canal and it is not always possible to scrape this area.

Technique

1. Place the patient in the left lateral or dorsal position.

2. With good illumination insert the speculum, lubricated only with water, and expose the cervix.

3. If the presence of an invasive carcinoma is suspected on clinical examination, refer the woman for a gynaecological opinion irrespective of the smear result.

4. Insert a wooden or plastic spatula into the cervical os and rotate through 360° (Fig. 15.3).

5. Spread the scrapings from the spatula thinly and evenly on to an indelibly named glass slide.

6. Fix immediately with a spray or by immersing the slide for 10–15 minutes in equal parts of absolute alcohol and ether.

7. Proceed to vaginal examination *after* carrying out the cervical smear.

8. Complete all sections of the request form, which must accompany each smear to the laboratory.

Table 15.4 Deaths from carcinoma of the cervix uteri: 3-yr periods 1968–76

			Age		All ages	
	25–34	35–44	45–54	55–64	65+	15+
Grampian region						
1968–70	4	7	15	24	28	78
1971–73	1	3	18	13	24	59
1974–76	—	5	12	12	23	52
Tayside region						
1968–70	1	6	16	22	17	62
1971–73	2	1	13	16	19	51
1974–76	—	4	9	11	27	51
All Scotland EXCLUDING Grampian and Tayside regions						
1968–70	10	49	165	164	223	611
1971–73	11	38	131	168	239	587
1974–76	14	41	109	180	221	565
England and Wales						
1968–70	98	601	1760	1850	2871	7194
1971–73	126	485	1632	1803	2723	6782
1974–76	176	386	1280	1790	2760	6417

Macgregor JE, Teper S 1978 Lancet ii: 774–776

Rotate spatula through 360°

Fig. 15.3. Taking a cervical smear.

Cervical intraepithelial neoplasia, carcinoma in situ

Exfoliative cytology identifies cellular abnormality but cannot diagnose the severity of the neoplastic change. Three stages of CIN are recognised (Fig. 15.4):

CIN I & II — equivalent to mild and moderate dysplasia.

CIN III — equivalent to severe dysplasia and carcinoma in situ.

Reversal to normal can occur in CIN I and II, especially in women under 30 years of age. Reversal to normal in cases of CIN III is uncommon.

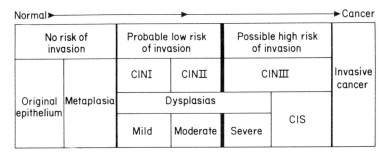

Fig. 15.4. The three stages of cervical intraepithelial neoplasia: CIN cervical intraepithelial neoplasia, CIS carcinoma in situ (Anderson M C, personal communication).

Management of women with abnormal smears

The *prevalence* of abnormal cervical smears is 3–7 per 1000 women screened, the *incidence* 1 per 1000 or less. An abnormal smear is not diagnostic of cervical cancer. It only identifies the presence of potential for disease and indicates the need for further evaluation. Normally the cytologist's report will contain suggestions for the appropriate management. If there is any doubt about this a personal approach to and discussion with the cytologist is the best plan.

In general

1. If the cellular abnormality is thought to be inflammatory in origin, treat the infection and repeat the smear three to four weeks after treatment is complete.

2. Mild dysplasia (CIN I) particularly in the postnatal and perimenopausal woman requires no immediate action. When the smear is repeated after six months the dysplasia has usually disappeared.

3. Smears reported as CIN I and II should be repeated after three to six months. If the abnormality remains after two smears, refer the patient to a gynaecologist for colposcopy.

4. Patients with CIN III smears should be referred to a gynaecologist. Wherever possible colposcopy is indicated. This allows detailed inspection of the cervix, the squamo-columnar junction and the vascular pattern. After staining with acetic acid the source of abnormal cells may be identified. Biopsy with direct colposcopic vision offers the best chance of an accurate histological diagnosis.

5. A patient who, after treatment, has two negative smears one year apart can enter the normal three- or five-yearly screening programme (see below).

Doctors should be clear about the facilities for investigation and treatment available locally. It is pointless to discuss theoretical possibilities with a patient. Where a colposcopic service exists, cone biopsy will be avoided in the majority of women with abnormal smears. Instead they will be treated by locally destructive techniques such as cryosurgery, laser vaporisation or cautery. Cone biopsy will still be necessary for the few patients in whom the abnormal epithelium extends up into the cervical canal and where its upper edge cannot be delineated.

Frequency of cervical smears

Every effort should be made to ensure that the maximum number of sexually active women are screened at regular intervals. Frequent repeat smears in women without risk factors are unnecessary.

The Committee on Gynaecological Cytology set up by the DHSS considered current policy on the age and frequency of screening for cervical cancer (Draper 1982). Its proposals are included in the following recommendations:

1. Virgins need not be screened for cervical cancer.

2. Cervical smears should be carried out in the sexually active woman:

    a. at 22 years if she has not had a previous smear;

    b. at 30 years if she has not had a smear during the previous 5 years;

    c. early in the course of each pregnancy;

    d. on at least one occasion between the ages of 22 and 35, if she asks for the test;

    e. at 5-yearly intervals after the age of 35 unless otherwise indicated. If women could be persuaded to have smears on their quinquennial birthday from 35 to 65 it would help them to remember when their next smear is due.

3. Women over 65 who have had at least two recent negative smears and have never had an abnormal smear do not require further cervical screening.

4. A diagnostic smear is indicated for any woman who has symptoms.

Although abnormal smears are uncommon under the age of 22, they have been reported in girls as young as 16. The above recommendations are therefore minimum requirements and, if the service and the laboratory can cope, the age range for smears should be expanded. Ideally one would like to offer a cervical smear:

1. At the beginning of sexual activity.

2. One year later to eliminate a false negative report.

3. Regularly thereafter with a maximum interval of 5 years.

4. More frequently to women at greater risk of cervical cancer — those who start having intercourse at an early age, who have multiple partners (or whose partners have multiple partners), who have a history of sexually transmitted diseases, particularly genital herpes or genital warts, and those who have previously had abnormal smears.

Remember. Preclinical cancers are now presenting at an earlier age (Fig. 15.5). Many women dying from cervical cancer are those who have never had a cervical smear.

The family planning consultation presents an opportunity for prevention of cervical cancer that must be taken up with enthusiasm.

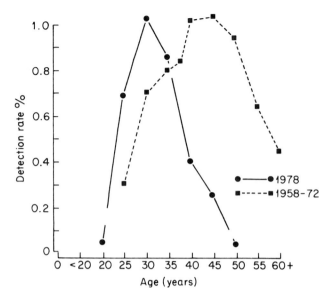

Fig. 15.5. Detection rate of preclinical cancers — now presenting at an earlier age, (Macgregor 1981). (●————●) 1978; (■-----■) 1958–72.

Cancer of the ovary and corpus uteri

These diseases are rare in women under the age of 45 (Table 15.1, Fig. 15.1). Routine screening to detect them is inappropriate for the majority of women attending a family planning clinic. However, every woman having a cervical smear should have a bimanual pelvic examination to identify uterine or ovarian enlargement. If a pelvic mass is found she should be referred to a gynaecologist.

If laboratory facilities are available, a smear may be taken from the posterior fornix in women over the age of 45 to detect abnormal endometrial cells which do not show up on a cervical smear.

Endometrial aspiration for cytological or histological study is not recommended in family planning clinics.

Other pelvic conditions

During routine pelvic examination nonmalignant conditions may be found. If the patient is told about their presence but not given adequate explanation, a great deal of unjustifiable anxiety can be caused. Doctors and nurses should be particularly careful to avoid this pitfall of screening.

1. Cervicitis, including retention cysts, Nabothian follicles, etc.

2. Cervical erosion (or cervical columnar ectopy). No treatment is indicated if the woman is symptom-free and has a normal cervical smear.

3. Cervical polyps require removal and histological examination. If symptoms are present a D & C may be necessary.

4. Ovarian cysts may be detected on routine bimanual examination or may present with lower abdominal pain as a result of torsion, rupture or haemorrhage. Ideally pelvic examination should be done on all new patients (it is necessary for those using a diaphragm or IUD), but should not be insisted on. There is no justification for performing regular yearly pelvic examinations in healthy young women just to identify ovarian tumours.

5. Uterine fibroids are uncommon in women attending for family planning, although they are slightly more common in young Negro than in Caucasian women.

6. Retroversion of the uterus, with or without retroflexion, is present in approximately 15% of women. If patients are told of this finding, they should be reassured that it is quite normal.

7. Sexually transmitted diseases.

a. *Gonorrhoea.* Although gonorrhoea is symptomless in 50% of affected females, routine screening for the disease is unnecessary at a family planning consultation since the pick-up rate is less than 0.5% of patients screened (Goldacre 1979). If the woman is known to be at high risk or has had intercourse with someone known to have, or suspected of having, the disease, screening at the appropriate clinic is recommended.

b. *Trichomonas vaginalis (TV).* Routine microbiological examinations to detect this organism are not necessary in symptom-free women. TV is never a commensal and, if its presence is identified on a cervical smear, both the patient and her partner should receive treatment.

c. *Candida albicans.* Yeasts are present in the normal vaginal flora, but their growth is kept under control by the acidity of the vagina and the normal vaginal bacterial flora. Routine screening to detect their presence is not helpful, but if the patient complains of symptoms, vaginal and cervical swabs should be taken and the appropriate treatment given. When *Candiada albicans* is reported on a cervical smear, treatment is not indicated unless the patient is complaining of symptoms.

SCREENING FOR BREAST CANCER

One in every 15 women will develop breast cancer, which causes 20% of all cancer deaths and is as common or commoner in the over 25s than *invasive* carcinoma of the cervix (Forrest 1980, Last 1981). It is only superceded in the 25–34 age group by carcinoma in situ of the cervix. (Table 15.1, Fig. 15.1). Only 1 in 10 lumps found in the breast is a cancer. Screening by physical methods is therefore a very nonspecific test and a large number of 'false positives' are found by routine breast palpation or self-examination. The only more sensitive test presently available to detect breast cancer is mammography, and this is not suitable as a routine screening procedure for asymptomatic young women attending a family planning clinic. A study is presently being undertaken by the DHSS to determine the value of mammography in the detection of breast cancer and its effect on long-term survival from the disease. Before such an expensive screening test is introduced it is proper that detailed evaluation should be undertaken, though the results may not be available for many years.

There is evidence that the earlier breast cancer is detected and adequately treated the longer the life expectancy, and it is to this end that screening is directed (Lancet 1984).

Screening techniques

Breast self-examination (BSE)

Every woman should be taught how to examine her own breasts and encouraged to do so each month thereafter, preferably just after the menstrual period. This is particularly important in women over the age of 35. Excellent leaflets are provided free by the Health Education Council and the Women's National Cancer Control Campaign. In addition many breast clinics and departments of surgery produce their own. Such leaflets should be available in family planning clinics and in doctors' surgeries (Appendix).

It is important to stress to every patient that if she detects an abnormality she should report at once to the clinic or her doctor and not wait until her next routine visit is due.

The fear of disfiguring treatment for breast cancer may explain the reluctance of many women to examine their own breasts or even to report breast lumps once they have been found. Less mutilating treatment is now available in many centres, and if women were told about this it might well encourage them to seek help as soon as they discover a lump in the breast.

Clinical examination

Breast examination should be carried out by the doctor or the nurse using the same thorough technique as is taught for breast self-examination, and should be offered annually to those at particular risk of breast cancer. They include women a) over 35 years; b) with a history of benign breast disease and biopsy; c) with a strong family history of breast cancer, especially in a first degree female relative; d) who have no children or had their first child after the age of 35; e) who have already had cancer of one breast.

Mammography

Although this is not suitable for screening young women attending for family planning, it may be offered to women with high risk indicators where facilities are available.

Management of the patient with a breast lump

Patients with multiple discrete lumps or a generalised increase in density in the breasts on clinical examination should be re-examined after the next period. Symmetrical bilateral lesions and those associated with bilateral tenderness often resolve after the onset of menstruation. A young woman with a breast lump which 'comes and goes' should not be referred *straight away* to a surgeon.

BUT – *any* woman with a breast lump which is constant, i.e. does not change in size at different times in the menstrual cycle must be referred to a specialist. Although only 1 in 10 lumps is malignant, it is important to identify that one case. There are no characteristics of the malignant lesion. Many malignant tumours present as smooth, mobile, well circumscribed lesions and it is important to recognise this fact. It is, however, most unlikely that well circumscribed lesions that have been present for several years will be found to be malignant.

OTHER SCREENING

Weight

Routine weighing is unnecessary. Overweight patients should have their weight checked and be advised about appropriate diet. Weight should be correlated to height and the woman given her ideal weight from national charts. Most women are weight conscious and need little encouragement to monitor their own weight. Care must be taken not to make a woman so embarrassed by weight gain that she stops coming for family planning advice. The Health Education Council's leaflet *Diet for Health* is of interest even to women who are not overweight and should be readily available in the clinic or surgery.

Blood pressure

As a screening base the blood pressure should be recorded on all new patients, but this ideal is rarely achieved. Even in routine medical examinations at school blood pressure may not be recorded. Renal disease and coarctation of the aorta are rare but certainly not unknown in completely fit young people. A baseline blood pressure is valuable before a patient becomes pregnant. This simple test, which in women not taking the pill need only be carried out once by the clinic sister, should therefore not be beyond the resources of every family planning clinic.

All women using hormonal contraception should have their blood pressure recorded regularly (Ch. 4).

Urine testing and haemoglobin estimation

These tests are not recommended routinely but should be carried out if indicated by symptoms or examination.

Rubella screening

Maternal rubella infection before 16 weeks gestation can lead to severe fetal abnormality. Safe efficient rubella vaccine is available and great efforts have been made to offer vaccination at school to all girls aged 13. Although this has proved very successful in some areas, the resource is greatly underused in others. In 1980 and 1981 uptake averaged only 84%. Information is accumulating both in this country and in the USA about the effectiveness of both rubella vaccination and its risks (Hinman 1983).

The family planning consultation offers an ideal opportunity to check the immune status of women before they embark on a pregnancy. A history of prior infection or vaccination is not significant in determining the immune status.

Recommended schedule

1. Nulliparous women should be offered a blood test to evaluate their immune status. If resources for this are not available women who give a history of rubella vaccination at school or previous rubella infection may be excluded. However, many such women will be rubella susceptible.

2. Arrange vaccination for rubella susceptible women either at the clinic, by the general practitioner or at a special community health clinic.

3. Ensure that no woman who is pregnant or suspected of being pregnant is vaccinated.

4. Warn the woman not to conceive within the next three months and provide her with efficient contraception to try to ensure that she does not do so. The importance of this has recently been questioned following evidence that vaccine virus rarely causes fetal abnormality (Hinman 1983).

5. If screening is carried out in a clinic or hospital, notify the general practitioner of the result.

6. Give the patient the result in writing. Ask her to keep it and take it to the antenatal clinic when and if appropriate.

Leaflets and posters about rubella are available from the Health Education Council, and should be displayed in family planning clinics and general practitioner surgeries. It is particularly important to give these leaflets to women who are shown to be rubella susceptible.

WHO SHOULD SCREEN?

Having considered the individual woman, the diseases for which she should be screened and the frequency at which screening procedures should be carried out, it is proper to consider who should undertake these examinations. All the tests described, including bimanual pelvic examination, cervical smear taking, instruction in breast examination and breast examination itself can be undertaken by a specially trained nurse. Nurses accept similar responsibilities in other fields of medicine such as in midwifery and in intensive care units. The screening procedures recommended are well within the scope of the trained nurse. Training of these nurses can be undertaken by the doctors with whom they work or by the appropriate hospital department.

HEALTH PROMOTION

Health is a very special commodity. It cannot be bought or sold, nor can it be forced upon people.

Smoking

This is one of the better known risks to health and the incidence of carcinoma of lung increases year by year (Table 15.3). Lung cancer will overtake breast cancer as the primary cause of cancer deaths in women if the present rate of increase is maintained (Table 15.2).

Ischaemic heart disease is not an important cause of death or disability in women under the age of 54 but in older women it is one of the main causes of mortality and morbidity. There is a very strong correlation between ischaemic heart disease and smoking.

Every encouragement should be given to women to give up smoking. After 10 years of nonsmoking, the risk of smoking-related disease is no greater than in the general population. This fact should be a great encouragement to women to give up the habit.

The family planning consultation is an appropriate time to offer advice. There are many aids for those who wish to give up the habit, and information about these should be available in clinics and surgeries. The Health Education Council and the Scottish Health Education Group have produced excellent literature on the subject, which should be available.

Exercise

Moderate exercise is beneficial. Most housewives reckon that they work very hard in the home, but this exercise is generally not sufficient to be of help. Exercise should be undertaken on a regular basis every day within the limits of tolerance and in a properly planned fashion. Such exercise should be planned to raise the pulse rate above 100 beats per minute and to sustain it at this level for 5–10 minutes. The clinic personnel should be able to recommend one or two suitable books or exercise tapes to help patients with their keep-fit programme.

Health promotion is a very general and gradual affair. It is important that information is reinforced without 'nagging'.

The family planning consultation offers an ideal opportunity for health care professionals to show an interest in all aspects of a woman's health and to offer her help and advice on how to stay healthy.

The reassurance which comes from finding no abnormality on

screening is important in improving the patient's confidence in her own good health. Women should be made aware of the difference between tests which are *medically indicated* e.g. taking the blood pressure of an oral contraceptive user, and *screening tests*, which are offered as an additional part of the family planning service. If patients do not wish to avail themselves of the latter, they can be omitted. However, it is a sad reflection on current health care resources that valuable screening tests are often omitted because of the shortage of manpower, time or space.

REFERENCES

Draper G J 1982 Screening for cervical cancer: revised policy. The recommendations of the DHSS committee on gynaecological cytology. Health Trends 1982 14: 37–40 (Reprinted British Journal of Family Planning 1982 8: 95–100)

Forrest A P M, Roberts M M 1980 Screening for breast cancer. British Journal of Hospital Medicine 23: 8–21

Goldacre M J, Watt B, Loudon N, Milne L J R, Loudon J D O, Vessey M P, 1979 Vaginal microbial flora in normal young women. British Medical Journal 1: 1450–1453

Hinman A R, Bart K J, Orenstein W A, Preblud S R 1983 Rational strategy for rubella vaccination. Lancet i: 39–41

Lancet 1984 Breast screening: new evidence. 1. 1217–1218

Last P A 1981 Breast screening and its evaluation. In: Shorthouse M A, Brush M G Gynaecology in Nursing Practice, Ch. 14 p. 173. Bailliere Tindall, London.

Macgregor J E 1982 Taking Uterine Cervical Smears. Published by the British Society for Clinical Cytology

Shapiro S, Venet W, Strax P, Venet L, Roeser R 1983 Ten to fourteen year effect of screening on breast cancer mortality. Journal of the National Cancer Institute 69: 349–355

Appendix

Instructions for self-examination of the breasts

1. Sit in front of a mirror stripped to the waist. Be sure you are sitting up straight. Look at your breasts carefully.

2. Is there any inequality between the size of your breasts? Has one breast recently become lower than the other?

3. Now look at the nipple area. Has one nipple become turned in? Is there any discharge? Always inspect the inside of your 'bra' for signs of a discharge. *Do not* squeeze the nipple or areola.

4. Now look at the skin of the breasts. Is there any puckering or dimpling? Is there any rash or change in skin texture? You may have to lift the breasts to see the under surface.

5. Raise your hands above your head and concentrate on the upper part of the breast that leads into the armpit. Is there any swelling or skin puckering?

6. Lean forward and examine each breast in turn. Is there any unusual change in outline, any puckering or dimpling of the skin or any retraction of the nipple?

7. Lie down in a relaxed and comfortable position. You may find it convenient to carry out this part of the examination with a soapy hand when having a bath.

8. First examine the left breast with the right hand. Use the front part of the flat of the hand, keeping the fingers together. It is important to learn how hard to press when examining the breasts; too hard will dull the sensation and too soft will not allow you to feel deeply enough. Never pinch the breast; if you do, you may feel lumps even in a normal breast.

9. Slide your hand over the breast, above the nipple, starting at the armpit and moving across to the centre of the body, pressing in to feel for lumps.

10. Repeat this action passing the hand from the outside inwards below the nipple.

11. Finally slide your hand across the nipple, making sure you have felt all parts of the breast.

12. Feel for lumps along the border of the pectoral muscles and in the armpits.

13. Now carry out items 8–12 with the left hand on the right breast.

(Printed by permission of BUPA Medical Centre, London)

Sexual problems and their management

RECOGNITION OF SEXUAL PROBLEMS

In the 1960s the so-called sexual revolution occurred. This coincided with the increased availability of more effective contraception, with increased 'liberation' for women and with the advent of a more relaxed attitude to sexual expression, influenced by both the media and pop culture. About this time Masters and Johnston, coauthors of *Human Sexual Response* were systematically studying sexual behaviour in laboratory and treatment settings, and women's magazines began more frank and open discussion about sexual behaviour and sexual problems.

Expectations about sexual behaviour underwent a fairly dramatic change, resulting in teenagers being put under increased pressure both by their peer group and the new pop culture to become sexually active. Adults, too, became more sexually aware and more likely to complain if their sexual relationship was not up to expectations.

Although many of these changes have reduced inhibitions and improved people's understanding of their own relationships and sexual responses, they have also meant that individuals are less prepared to tolerate sexual problems, or accept mediocrity within their relationships. Marriages in Britain are now breaking down at a rate of one in four. Many changing social factors have been blamed for this increase. Dissatisfaction with the marital sexual relationship is one.

People generally turn to the medical profession for help with sexual problems, although many of them may experience difficulty in broaching this 'embarrassing subject' with a doctor. Some feel more comfortable with a doctor they know. Others prefer to seek help from a stranger either within their own general practice or in a family planning clinic setting. It is particularly important that both general practitioners and family planning doctors should be on the alert for any hint that the patient wishes to talk about a sexual problem.

Opening up the subject of sex can be embarrassing and difficult not only for patients but also for doctors, who can feel awkward and inept and may consciously or subconsciously discourage patients from disclosing a sexual problem. Many have received little or no training in the recognition and treatment of such problems and, more important still, have never had the opportunity to question their own personal sexual attitudes, values and prejudices which is so necessary when learning to handle the subject in a sensitive yet professional manner.

This chapter briefly outlines an approach to the recognition, classification and simple treatment of sexual problems encountered in the clinical setting. First, it is important to understand normal sexual response.

NORMAL SEXUAL RESPONSE

This can be divided into five phases as shown in Fig. 16.1 below:

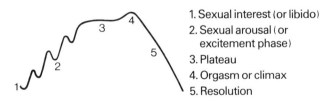

1. Sexual interest (or libido)
2. Sexual arousal (or excitement phase)
3. Plateau
4. Orgasm or climax
5. Resolution

Fig. 16.1

Sexual interest

Also called sexual appetite, drive or libido.

1. Is highly sensitive to variations in the couple's circumstances and changes in the emotional climate of the relationship.

2. Varies from person to person at different times, shows marked variation during the menstrual cycle in many women and decreases naturally with age.

3. Is maximal in men aged 15–35 and in women aged 25–45.

4. Is directly related to frequency of intercourse so that infrequent sexual contacts lead to a reduced interest in sex in most people. Average frequency of intercourse among married couples is 2.4 times per week but there is much variation within this norm.

Sexual arousal

1. Occurs from anticipation of sexual contact, from fantasy or, more usually, from physical stimulation such as kissing, body caressing or sexual intercourse.

2. In the woman it produces a subjective feeling of excitement and arousal and physiologically is manifested by vascular engorgement of the vagina, particularly the lower third, and a transudate of clear mucus from the epithelium of the upper two-thirds. The upper vagina balloons outwards and the cervix is retracted into the abdomen as sexual arousal increases.

3. The man also feels subjectively aroused and excited. Increased blood flow into the penis and closure of venous valves produce an erection. This often happens very rapidly in young men.

4. Feelings of arousal may show peaks and troughs in a perfectly normal response pattern in both men and women. In the man these fluctuations in the level of arousal may be reflected in variations in the firmness of his erection.

Plateau

If effective sexual stimulation is continued, sexual tension is intensified and subsequently reaches an extreme level which may lead to orgasm.

Orgasm

Also called climax.

1. In a woman this is the sudden release of sexual tension associated with rhythmic contraction of the pelvic muscles and a feeling of well-

being and exhilaration. Many women do not experience an orgasm until well into their twenties. By this time their self-confidence has increased and they are generally feeling more relaxed and trusting within a more stable sexual relationship.

2. Approximately 40% of women learn to understand their own body's sexual response by masturbation and will already be familiar with the experience of orgasm before entering into a sexual relationship.

3. As many as 80% of women need clitoral caressing as well as vaginal intercourse to reach orgasm during love-making. This is often not recognised by young couples who may presume that they have a sexual problem because the woman is not vaginally orgasmic.

4. In men, orgasm also signals a release of sexual tension usually associated with rhythmic contraction of the perineal muscles and prostate and ejaculation of semen.

5. Most young men masturbate in their teens and learn to experience arousal and orgasm in this way. Many inexperienced youngsters ejaculate extremely rapidly when sexually aroused. In early sexual relationships they may ejaculate either before, or shortly after, vaginal entry, causing great embarrassment to themselves and sexual dissatisfaction in their partners. This premature ejaculation is well within normal limits to begin with and most men quickly learn to delay ejaculation in order to enhance their own and their partner's pleasure in love-making.

Resolution

1. Once a man has ejaculated he enters a refractory period during which further arousal and erection cannot occur. In a young man this may last for a few minutes but in an older man the refractory period lengthens and may last up to several hours.

2. Women do not have such a clear refractory period and some are capable of returning to high levels of sexual arousal and further orgasm very quickly.

COMMON DISORDERS ENCOUNTERED IN A FAMILY PLANNING INTERVIEW

A survey of attenders at a family planning clinic in Edinburgh (Dickerson et al 1976) showed that at least 12% of sexually active women (mean age 28) felt that they had a significant sexual problem. This suggests that the percentage of sexual problems in the population at large may be much greater.

It must be emphasised that sexual problems exist between two people. There is no such thing as an uninvolved partner. Often both partners have some degree of sexual difficulty by the time they seek help. For example, a woman complaining of unresponsiveness often has a partner with premature ejaculation, although the couple may not reveal this at first. It is often impossible to determine which is the primary problem. In fact this is of little consequence as each is now maintaining the other's problem.

The commonest sexual problems seen in general practice and family planning clinics are sexual dysfunctions.

Sexual dysfunctions

These may affect anyone, in any kind of relationship. The principles for the management of such problems in homosexual relationships are no different to those for heterosexual relationships. The most common sexual dysfunctions are listed in Table 16.1.

Table 16.1 Sexual dysfunctions

	Disorders of interest	Disorders of arousal	Disorders of climax
Male	Low libido	Erectile problems, (impotence)	Premature ejaculation Delayed ejaculation
Female	Low libido	Unresponsiveness, (frigidity)	Preorgasmic problems Nonorgasmic problems
		Vaginismus	

Disorders of interest

Low libido. This problem is common in both men and women. Since the man is usually the initiator of sexual behaviour, low libido in the male is the more likely to affect the frequency of intercourse in a relationship.

A woman often loses interest in sex after childbirth. A variety of factors may be responsible including fatigue, hormonal effects, fear of another pregnancy or the woman focusing more exclusively on her parenting role at the expense of her role as a wife.

Approximately 10% of women taking the contraceptive pill complain of reduced sexual interest and enjoyment. A similar number report enhanced sexual experiences. There is strong evidence that this is subjective, not hormonal.

A marked discrepancy between partners' natural libido may lead to sexual difficulties as the one with less interest feels continually under sexual pressure and the other resentful of wasted opportunities.

In both sexes anger, anxiety, long-standing inhibitions, chronic fatigue, overwork and physical illness can have a marked effect in reducing libido.

Disorders of arousal

Male erectile problems. These are the most obvious and devastating of all sexual problems. They represent a threat to a man's masculinity and self confidence.

Erectile failure may be transient, intermittent, primary or secondary and is often situational, e.g. he can get an erection by masturbation but not in the context of a relationship.

Physical causes such as diabetes, neurological diseases, alcohol abuse, drug treatment (especially for hypertension), or local pathology must be excluded, by careful history-taking and appropriate examinations.

Psychological rather than physical causes are commonly responsible for erectile failure, e.g. following myocardial infarction, after being made redundant or any such personal crisis which shakes a man's self-confidence.

Unresponsiveness in the woman. This is less obvious than impotence. Although vaginal lubrication may be reduced, the woman can still make love. However, intercourse may be uncomfortable or even painful as the result of vaginal dryness and penile pressure on the cervix, which is low lying in the pelvis if she is unaroused. She may therefore present with dyspareunia. She may find intercourse distasteful and often reports feeling 'dead from the neck down'.

Physical causes include vaginal infections such as moniliasis, senile vaginitis and postmenopausal vaginal changes.

Pathological causes of pelvic pain must be excluded before dyspareunia can be attributed to unresponsiveness.

Female unresponsiveness is often secondary to premature ejaculation in the partner.

Unresponsiveness in the female may lead to erectile problems in her partner if he is sexually vulnerable and unsure of himself.

Disorders of climax

Premature ejaculation. Ejaculation occurs sooner than either partner would choose and may be before vaginal entry or very soon after intercourse has started.

It is not unusual and can be considered normal in young or sexually inexperienced men. If, however, the man is unable to learn to delay the time of his ejaculation he is likely to become sexually anxious and so further exacerbate the tendency to premature ejaculation. Furthermore, his partner may have lost interest in sex which reduces the frequency of intercourse and makes the problem worse.

Many couples will not admit to this problem, the wife because of loyalty to her husband, the husband because he is ashamed of, or embarrassed by, the difficulty.

Delayed ejaculation. This is a less common problem but causes considerable distress particularly if a pregnancy is desired. It is usually a sign of sexual inhibition of a long-standing nature. Some men can ejaculate with masturbation but not intravaginally.

Orgasmic problems in the woman. Many women do not experience orgasm each time they make love. Preorgasmic women have never experienced orgasm. They may fear the unknown and build up all sorts of fantasies about what an orgasm may mean such as loss of control and dignity, incontinence or pregnancy. Others have more deep-seated psychological reasons for being afraid to 'let go' and relax with their partner. Faulty love-making technique such as inadequate clitoral stimulation is a common cause of orgasmic difficulties.

Anxiety and tension tend to delay a woman's climax while they accelerate a man's.

Vaginismus

This is specifically a female sexual problem and does not fit into any classification very satisfactorily. Marriage may be unconsummated and the woman will often describe how she is interested in sex, enjoys sexual intimacy, is capable of arousal and even orgasm yet resists any attempt at penetration of the vagina by her partner's penis or even his finger.

Spasm, often severe, of the vaginal muscles (pubococcygeus) and the adductor muscles of the thighs develops. This may simply be due to fear of the unknown, ignorance of her anatomy or fear of being hurt. It might be the result of a more deep-seated anxiety, based on a past unpleasant sexual experience such as rape, incest or physical trauma. Serious emotional hurt, perhaps delivered by a male in the past, may be the cause.

Some women develop vaginismus as a defence mechanism against growing up and becoming a woman. These women often live close to their parents and see their role more as a daughter than as a wife and potential mother. They resist sexual intercourse because it is the symbolic act of becoming a woman and frequently marry unassertive men who do not complain about the nonconsummated marriage, being sexually anxious and unsure themselves. Although the woman may wish her partner to be more assertive and even force his way with her sexually, she still feels safer with his unassertiveness and her 'doll's house' marriage.

Sexual deviations

Sexual deviations, e.g. homosexuality, trans-sexualism, transvestism and fetishism are much less common and are outwith the scope of this book.

CAUSES OF SEXUAL DYSFUNCTIONS

Many factors lead to sexual dysfunction in a couple (Table 16.2). Sexual problems affect both partners and once a problem has become established both tend to maintain it by developing performance anxiety.

Table 16.2 Causes of sexual dysfunctions

Physical	Psychological
Endocrine disorders	Ignorance and misunderstanding
Neurological diseases	Anger and resentment
Chronic illness	Anxiety and guilt
Drug effect	Poor self-image
Fatigue	'Commonsense' factors
Alcohol	
Pelvic pathology	

'Performance anxiety' means that the couple are no longer relaxed and spontaneous in their love-making but, alert to further failure, become detached spectators to their own sexual performance, watching out for further erectile failure, premature ejaculation or lack of orgasm. Performance anxiety is present in every couple complaining of a sexual problem. Although probably not responsible for the original problem it must be considered as a factor in maintaining it.

A simple classification of common sexual dysfunctions is shown in Table 16.2. It must, however, be remembered that complex patterns of interactions are hidden within this table and that a specific psychological or physical problem may lead to a different dysfunction in different couples. For example, anxiety and anger may lead to erectile problems in one couple and orgasmic problems in another. Physical and psychological problems often interact with each other.

Physical problems

Physical problems such as diabetes and other endocrine problems, neurological diseases, chronic illness, drug effects, fatigue and alcohol may be the cause but are not common.

Local causes include pelvic pathology such as vaginal infections.

Psychological problems

Ignorance and misunderstanding

Ignorance and misunderstanding about sex often lead to problems. Faulty sex education, faulty sexual technique, inability to communicate sexual needs to each other and unrealistic expectations about sex may all be to blame.

The stereotyped views about sex which some people have, e.g. men always initiate, women should remain passive and keep control, may be contributory factors.

Simple problems such as ignorance and sexual anxiety are highly amenable to simple counselling which should be undertaken almost routinely with young and inexperienced married couples. The family planning consultation is an ideal time to check simple factors which could cause sexual difficulties.

Anger and resentment

Anger or chronic anxiety about being hurt or deceived can often remain unresolved, because the couple cannot communicate their feelings clearly to each other or have a row to clear the air. Each complains that their partner has not understood their feelings, yet at the same time is not prepared to listen to the other's point of view.

Common problems include the bringing up of children, finance, involvement of relatives in the marriage, intrusive hobbies, drinking habits or a power struggle between the sexes. All these factors can create resentment which in turn can lead to sexual difficulties within the marriage.

Extramarital relationships often have a devastating effect, leading to loss of trust and self confidence.

Anxiety and guilt

Negative sexual attitudes and values are often laid down in childhood by a strict parental upbringing which looks upon sexuality and any form of sensual pleasure as undesirable and bad. Guilt feelings and inhibitions thus developed in childhood are often difficult to overcome in adulthood. Traumatic sexual experiences add to these fears and anxieties.

Violence in the home in childhood may lead to permanent distrust of intimate relationships in adult life.

Past history of incest or rape often leaves a person with chronic sexual anxiety and disgust, leading to deep-seated sexual problems.

Poor self-image

This may be caused by problems of body image or self-esteem. Women can be 'put off' sex by chronic anxiety about their breast size, the size of the labia minora etc., while men worry about such things as the size of the penis. Both may be affected by feelings of sexual inadequacy or general unattractiveness perhaps because of a physical handicap. Without doubt, if one feels unattractive one often becomes uninterested in sex and incapable of responding.

After a mastectomy or hysterectomy or even after the menopause, many women develop sexual problems related to their altered self-image. Depression undoubtedly affects a person's self-esteem and self-image.

Many women, housebound with small children, report feeling a loss of interest in sex which is again tied up with their own feelings of low self-esteem. Men who are made redundant develop similar difficulties.

'Commonsense' factors

Fear of pregnancy, fear of discovery by adults, fear of being overheard by teenage children, the baby in the same bedroom or the mother-in-law next door are all potent factors in producing an impaired sexual response.

ASSESSMENT OF SEXUAL PROBLEMS

The family planning consultation provides an ideal opportunity to enquire about physical or sexual relationships. Patients often feel inhibited or anxious about raising the subject themselves, but may drop hints or even deflect sex-related psychological symptoms into physical manifestations, such as chronic backache, mild urinary symptoms or vaginal discharge.

If the patient does acknowledge some difficulty in this part of his or her life the doctor should enquire further into the subject thus giving the individual permission to describe the problem in more depth. Such 'closed' questions as 'Does either of you have trouble in reaching a climax?' 'Do you have sufficient lubrication?' 'Does your partner come too soon?' help the patient to use sexually-explicit language, but the doctor must ensure that his vocabulary is clearly understood by the patient and vice versa. He must make it quite clear that he is not embarrassed by talking about the subject, and that it is appropriate for the patient to talk about it too.

The doctor must then decide whether or not the patient wants to pursue the problem further. If the patient wishes further help the doctor should encourage the patient to talk more, by asking 'open' questions, e.g. 'How does this problem affect your relationship?' 'What do you feel about it?'

Sexual history taking

Because sexual problems are caused by many factors it is usually impossible to discover all the complex interactions which may be occurring. The following areas should be explored:

1. Does the sexual problem interfere with the sexual behaviour of the couple?

2. When did it begin?

3. Was it associated with any other problems in their lives at that time — in the relationship or within themselves?

4. What effect has the sexual problem had on their relationship?

5. How often do they have intercourse now?

6. Is the problem present when they masturbate? This question should always be asked alone and not in front of the partner.

7. How long have they been married; the number of children; housing conditions; work role, etc?

8. Significant past events in personal relationships including upbringing, etc.

9. Recent physical illnesses and treatment.

10. Method of contraception.

Examination

In most cases a full physical examination is unnecessary unless specific symptoms in the history suggest otherwise. Situational sexual problems (e.g. not present with masturbation) are rarely organic. Examination of external genitalia and vaginal examination should always be undertaken to exclude local pathology, to aid diagnostic assessment of vaginismus, to reassure, and to educate the patient. Urinalysis should be routine (especially in unexplained erectile failure) to exclude diabetes.

TREATMENT OF SEXUAL PROBLEMS

The same principles of simple, sexual counselling apply regardless of the particular form that the sexual problem takes.

Physical problems

Physical problems must first be excluded or dealt with if present. A commonsense and practical approach will need to be adopted, to assist the couple to come to terms with physical problems such as a chronic handicap, which cannot be resolved. This may involve suggesting alternative positions for intercourse, giving permission for the couple to enjoy caressing to orgasm without vaginal intercourse and the use of sex aids such as KY jelly and vibrators, where needed.

This topic is dealt with by Bancroft in *Human Sexuality and its Problems*.

Psychological problems

Once physical causes have been excluded, sexual counselling should be directed towards the psychological factors which are behind most sexual problems. If at all possible the clinician should work with the couple rather than with one partner, as sexual problems invariably involve problems for both. If one partner is unwilling to attend for counselling then the following principles may have to be modified for use with an individual.

As all sexual problems are usually caused by a mixture of background factors, several objectives are essential when counselling patients with sexual problems. These may be summarised as follows.

Reduction of ignorance and misunderstanding

Even a couple married for several years can benefit from being reminded of the following points:

1. Sexual responses are a natural phenomenon and occur when an individual is relaxed, wanting to respond, and receiving appropriate stimulation.

2. There is no 'correct' sexual behaviour. Each couple must work out for themselves what seems comfortable and acceptable.

3. Each should show what they do and do not like when being caressed and should encourage their partner to do the same. This applies especially to genital touching.

4. Despite the man's early erection both partners benefit from a period of foreplay (kissing and caressing) before intercourse.

5. Premature ejaculation is common, but can be delayed with practice and experience (see specific treatment).

6. Most women reach a climax more readily with stimulation of the clitoris prior to or coincident with vaginal intercourse.

7. It is not necessary to aim for simultaneous climax as this often produces 'performance anxiety' and it can be just as pleasant to experience each other's orgasms separately.

8. Both partners should take responsibility for initiating sex and both should feel able to say 'no' without fear of an angry response. Similarly if they say 'yes' it should have no conditions attached.

The clinician may wish to recommend a couple to read together such books as *The Joy of Sex*, *Treat Yourself to Sex*, or *Coping with Sexual Relationships*.

Reduction of anger and resentment

Unresolved resentment is frequently present despite initial protests from the couple that they never argue and have an ideal marriage apart from the sexual problem. The following steps can enhance and clarify marital communication. Encourage them:

1. To express their own feelings by concentration on self-assertion — 'I would like to do ... ', and self-protection — 'I feel hurt when ... '. This is difficult for most couples who regularly criticise each other instead — 'If only you would ... ' or 'Why don't you ... '.

2. To listen to their spouse as well as concentrating on their own needs and feelings.

3. To begin to negotiate fairly to produce a mutually acceptable approach to chronic problems such as finance, etc., once they have

heard each other's needs and what upsets them. Each should be gaining and giving concessions. This process may need to be practised with a referee at first in the surgery and then transferred to the domestic setting.

4. To praise rather than criticise. They should aim at noticing and remarking on something nice done by their partner every day.

5. To spend at least 10 minutes alone together talking about their day and their feelings, each being given the same amount of time and interest by the other.

Reduction of anxiety and guilt

1. Performance anxiety. For a short time a couple can be encouraged to set limits in their lovemaking to avoid goal-orientated sex. This helps especially those with orgasmic problems and erectile difficulties. An agreed contract to ban intercourse and to concentrate on mutually pleasing caressing should be made, with the emphasis on doing both what each individual enjoys and what each knows the partner likes. They should be instructed to avoid trying to get a response and to concentrate instead on relaxing and enjoying the sensations being given and received. Any unilateral breach of the contract (by one partner insisting on intercourse) should be taken seriously as violating the couple's mutual trust and sabotaging the treatment approach.

2. Anxiety related to fears about sex. Simple fears about pregnancy, loss of control or incontinence associated with orgasm should be allowed expression and taken seriously. Appropriate counselling by the clinician using his medical authority and knowledge is then important. More deep-seated feelings of guilt, distaste and anxiety about sex, based on attitudes and values developed in childhood, must also be tackled. The doctor's more permissive attitude to sex may help patients to question their clinging to infantile beliefs and encourage them to dismiss taboos and inhibitions from their past. Showing a couple explicit pictures of normal lovemaking or sex education films, may help break down inhibitions and prejudices.

3. Mistrust of sex. Mistrust of sex or the opposite sex will need to be aired with the partner and every encouragement given to help increase the security of the relationship, by frank communication between them. Individuals in whom inhibitions and fears prove resistant to simple counselling may need referral to a specialist psychiatric or psychosexual agency.

4. Tranquillisers. A prescription for tranquillisers may on rare occasions be helpful; they should be taken 1–2 hours before sexual intercourse.

Improving an individual's self-image

An understanding counsellor can provide the support and encouragement which these patients need, in the following ways:

1. The doctor must beware of talking down to a depressed or self-deprecating patient thus enhancing their feelings of dependancy and poor self-confidence. These feelings should be respected as understandable and every effort made to persuade the patient to find ways of boosting morale. The doctor's respect will start this process.

2. Social contacts, hobbies and pride in work role to promote reinforcement from social relationships should be encouraged.

3. Particular worries such as postoperative disfigurement and breast size should be discussed with the couple. The person's feelings should be respected by both the doctor and the partner rather than made light of. They should empathise and at the same time be reassuring about their own attitudes to the offending problem where possible.

4. Occasionally antidepressants may be indicated.

Specific treatment techniques

Premature ejaculation

In addition to the above counselling, the couple should use the stop/start technique to delay the time of ejaculation. During caressing or intercourse when the man feels that he is close to a climax — that he has almost reached the point of no return — he should stop being sexually stimulated, relax for 30 seconds, then recommence whatever he was enjoying until he gets close to climax again. He then repeats the relaxation period. He is bound to mistime initially but, with practice, he can educate his body to delay the ejaculatory reflex. Once confidence increases, the learning is rapidly consoldiated. If this is insufficient he can add the 'squeeze technique'. This involves firmly squeezing the tip of the penis between finger and thumb during the relaxation period.

Delayed ejaculation

In addition to working with the anxiety and guilt which usually inhibit normal ejaculation, a ban should be put on intercourse and the couple encouraged to mutually caress, both keeping as relaxed as possible. The woman should caress the penis to produce maximal stimulation and sensation while the man concentrates on remaining relaxed. If performance anxiety develops and he feels tense, the couple should stop for a while and then resume contact when he feels relaxed again. He should also be encouraged to masturbate to climax alone using fantasies about intercourse with his wife. If this fails, referral to a specialist should be suggested.

Orgasmic dysfunction

1. Preorgasmic women may need permission to learn about their body's responses by self-stimulation either from masturbation or from using vibrators. Once the woman has learned to relax enough to enjoy herself and to allow herself to reach a climax, she should be encouraged to show her partner how she enjoys being caressed and incorporate this into their normal lovemaking.

2. For women who are nonorgasmic in their marital relationships, it is usually essential to apply the general principles of sexual counselling already described, with particular emphasis on lack of trust, anxiety and poor self-image. Several booklets on self-help for women are available.

Vaginismus

Gradual desensitisation of the woman's fears of vaginal entry may be all that is needed in simple cases. This can be done by sympathetic vaginal examination, by encouraging the woman to examine herself or use a vaginal dilator and also by encouraging the husband to become confident enough to insert his own finger into her vagina. If he is totally inexperienced be may be taught to do this in the clinic at first.

If a programme of graded dilator use and encouragement fails, it may be that the couple have more deep-seated anxieties about consummation. The wife may need help with her adult identity and separation from her parents. In questioning her real motivation to grow up she is confronted with the problem and this may clarify whether or not she feels emotionally ready to leave home. Difficult cases may require referral to a specialist.

WHERE AND WHEN TO REFER

A couple should be referred to a specialist agency when their problems cannot be solved by simple counselling or where deep-seated anxieties and marital problems are predicted.

Psychosexual clinics staffed by trained counsellors from a variety of disciplines are held in many areas. These clinics may be run by psychiatrists, psychologists, gynaecologists, family planning doctors or marriage guidance counsellors.

If no facilities for psychosexual counselling exist in an area, a small group of doctors should be encouraged to meet regularly as peer group support while group members train themselves, using the modified Masters and Johnson approach to psychosexual counselling, as outlined in their book *Human Sexual Inadequacy*. Such groups need not consist of doctors alone but may be multidisciplinary.

CONTINUATION OF SUPPORT

Sexual problems may arise at any time in a couple's relationship with particular crisis points being in early marriage, at the time of childbirth and the menopause, during illness and after surgery or bereavement. The clinician is frequently consulted during these crises and should enquire about sexual problems. Brief intervention at an early stage can often prevent a long-lasting sexual difficulty from developing.

Patients with minor problems can be followed up during routine consultations but, for counselling of those with an established sexual problem to be effective, regular and specific follow up of the couple is crucial. 4 sessions of 20–30 minutes each will be sufficient to produce a significant change in attitude, anxiety and ignorance. It will be time well spent.

Although routine enquiry about sexual problems is to be encouraged, particularly in the context of contraceptive practice, care must be taken not to intrude when patients clearly indicate that they do not wish to discuss this private part of their lives with the doctor. There is a thin dividing line between clinical zeal and the right to privacy.

REFERENCES

Begg A, Dickerson M, Loudon N B 1976 Frequency of self-reported sexual problems in a family planning clinic. British Journal of Family Planning 2: 41–51
Masters W H, Johnson V E 1966 Human sexual response. Little Brown & Co,
Masters W H, Johnson V E 1970 Human sexual inadequacy. J & A Churchill, London

FURTHER READING

Bancroft J 1983 Human sexuality and its problems. Churchill Livingstone, Edinburgh

Brown P, Faulder C 1980 Treat yourself to sex: A guide for good loving. Penguin

Comfort A 1983 The joy of sex. Mitchell Beazley (New edition)

Fairburn C G, Dickerson M, Greenwood J 1983 Sexual problems and their management. Churchill Livingstone, Edinburgh

Greenwood J 1984 Coping with sexual relationships. MacDonald, Edinburgh

17 *George M. Morris* & Alistair J. Moulds*

Family planning in general practice

Giving advice on family planning and contraception involves the general practitioner in health education and practical preventive medicine at its best. The large numbers of unwanted pregnancies that still occur in Great Britain leave no room for complacency over the effectiveness of our family planning programme. There is certainly room for a greater contribution from well-trained, enthusiastic general practitioners, some of whom are still not taking advantage of their unique position to provide a service for the most poorly motivated patients.

This service can be given in many ways. Some practices may be able to support an IUD clinic and fit caps as well as prescribing the pill and giving general advice. Within smaller practices, including single-handed ones, an excellent service can be given. This will usually consist of general advice, prescription of the pill and referral to clinics, health centres or hospital for the provision of other services. Some general practitioners run separate family planning clinics but most doctors prefer to give contraceptive advice at the same time as general medical services.

Each method has its advantages and disadvantages. The important thing is that the doctor must adopt a flexible approach. The woman who needs a pill prescription urgently or postcoital contraception must not be kept waiting for several days. Sometimes the need for contraception is identified at a general medical consultation during a busy surgery. The doctor may then think it right to take a briefer case history than usual and, if the pill is chosen, may confine the

*Deceased 1984

313

examination to taking the blood pressure. It is better to provide immediate contraceptive cover in this way than to keep the patient waiting for an appointment at a family planning clinic.

Although most of the general practitioner's advice will be to women, this is by no means always the case, and the doctor must remember that male patients may want help also even if they do not ask for it.

PRESENT POSITION

Free family planning became available to all, irrespective of age and marital status, at hospital and community clinics in April 1974 and from general practitioners in July 1975. Nearly 97% of family doctors now provide contraceptive services, although they are not obliged to do so, and approximately 2½ million women receive contraceptive advice from them. These figures are impressive but there is still scope for improvement.

First, some general practitioners merely prescribe the pill, rather than provide a full contraceptive service. Only 20% fit intrauterine devices.

Second, there is great potential for increasing the number of patients being counselled. In a practice 20% of patients are females in their reproductive years and it is likely that most will, at some time, require or would benefit from advice on contraception. As this group consults, on an average, five times each year, there are many opportunities to discuss and review their contraceptive needs. The general practitioner is in an excellent position to give advice and help that may benefit the patient and her family in years to come.

It should now be routine to discuss family planning at the postnatal visit, at the time of termination of pregnancy, on the numerous occasions when the woman herself consults and even possibly when she brings her children to the surgery. At some consultations, of course, it would be entirely inappropriate to broach this subject although it is surprising how often it is not. With a systematic approach, contraceptive advice can be offered tactfully without any problem and without using too much valuable time.

Actively seeking out patients is truly worthwhile. Once those in need of advice are identified it is important that the general practitioner has a comprehensive and up-to-date service to offer them either inside or outside the practice. This should be of such a standard as to give the patient both confidence and satisfaction. Whatever doctors may think about method safety, many patients do have fears and reservations. The knowledge that they are being well looked after by their doctor often allays these fears and reassures them. On the other hand, an overactive approach may deter patients from seeking or accepting contraceptive advice. This is particularly true in relation to vaginal examination and taking cervical smears. The doctor must assess each case on its merit and adopt an understanding and flexible approach.

Under National Health Service regulations general practitioners have an obligation to provide general medical services for all patients on their list. They may also, if they wish, provide contraceptive services not only to their own patients but also to those not registered with them.

Family planning payments are now incorporated in their general remuneration but have to be claimed as a fee for item of service (Appendix). The form has to be signed by both patient and doctor and covers contraceptive services for 12 months from the date of acceptance.

Women are entitled to seek contraceptive advice from their general practitioner and from a clinic at the same time. Attendance at a clinic does not preclude the general practitioner from giving contraceptive treatment or advice or from claiming payment.

PROVIDING A FAMILY PLANNING SERVICE IN PRACTICE

The role of the general practitioner

Work from a sound clinical base

The doctor must have all the necessary background knowledge clearly sorted out in his or her own mind otherwise the relevant information cannot be successfully communicated to patients who now expect this. Understanding and compliance are more likely when the doctor has taken the trouble to explain the reasoning behind the advice given and the patient has been fully involved in the decision making.

Knowledge has to be regularly updated with relevant reading and attendance at appropriate meetings and courses. It is important that doctors should know about new advances in contraception and new risks that are being identified before their patients hear of media reports, which may not always be accurate.

Adopt a positive approach

Ideally, every patient of reproductive age should have a clear record of her current need for or method of contraception entered in her notes.

The doctor can then tactfully ask these patients about their contraceptive needs and even discuss other options that may be open to them. Some situations, e.g. patients registering with a practice for the first time, women with genito-urinary symptoms, young women consulting with apparently minor problems are more 'fertile' ground for this sort of enquiry than others. As long as the doctor is sensitive and obviously concerned with helping them, no offence is likely to be given or taken.

Once in the picture it is important to keep a sense of proportion in deciding what to do with the information. If couples are realistically and happily using methods such as coitus interruptus or natural family planning, it may be advisable merely to give advice on how these methods may be used more effectively.

Provide access to a comprehensive service

In some practices at least one member of the practice team is able to fit IUDs and diaphragms. Very occasionally a vasectomy service can be offered as well. If this is not the case, referral should be made to a community family planning clinic, a hospital or to another practice where a greater range of services is provided. Referral may also be considered appropriate when a specialist opinion is required. This could be when medical conditions make the finding of an effective contraceptive difficult. It could also be where social and legal difficulties are likely to arise, as with the very young or with mentally subnormal patients. There should be no need for patients to be referred elsewhere for routine screening procedures.

It is important to ensure that patients know that a comprehensive service is being provided. Word of mouth recommendations certainly help but a clearly written notice prominently displayed in the waiting room and a leaflet for new patients are even more effective. In particular, they draw to the attention of young people their doctor's attitude to the provision of contraceptive care.

Aim for a consistent approach throughout the practice

Ideally all members of the team will be in general agreement with the practice policy, have the same attitude to patient's requests and present approximately the same facts. One woman can quickly undermine a neighbour's confidence in her method or in her doctor by quoting conflicting advice or information which she has received from a partner. The standard and type of care a woman receives from the practice should not depend on which partner she happens to consult.

To this end it is very helpful if all partners can agree a policy for the provision of contraceptive care and screening within the practice. However, where one member of the team is at odds with the attitude or ideas of the others, it should be possible, given good will, to cope with the situation without detriment to the patient.

The role of the other members of the primary care team

The entire practice team has a contribution to make towards an efficient family planning service. Secretaries, receptionists and telephonists should be able to advise patients of the facilities provided and how to gain access to them.

Midwives, health visitors and community nurses should be aware of the particular needs of the diverse groups of patients with whom they come into contact — married couples, the very young, patients with language difficulties, the socially disadvantaged, the mentally and physically handicapped, etc. — and should feel able to take the initiative both in discussing contraception and in arranging consultations for those reluctant to do this themselves. They can also help provide a domiciliary service for those in need of it.

Practice nurses who have undertaken the appropriate training (Ch. 1) extend the range of the service provided within the practice since they are capable of accepting a high level of clinical responsibility and can undertake much of the doctor's routine family planning workload. A well-trained and experienced practice nurse can fit and teach the use of the diaphragm and see patients at their follow-up visits. She can also see patients on the pill at routine return visits to check that all is well before the doctor signs the prescription. Some doctors are confident enough to allow the nurse to undertake routine checks in patients who have been fitted with the IUD. Nurses are also able to take cervical smears and to teach self-examination of the breasts. The nurse can also provide time and a sympathetic ear for those who need it. The extra time that she can spare and the fact that she is a nurse rather than a doctor, may allow some patients to discuss more freely with her problems they consider either too minor or too intimate to bring to the 'busy' doctor's attention.

The extension of family planning responsibilities to other members of the primary care team brings obvious advantages. However, the more people who are aware of confidential information relating to a patient's contraception and personal problems the greater is the risk of confidentiality being breached. The doctor has strong legal and moral responsibilities to ensure that no breach of confidentiality occurs and that all respect the patient's fundamental rights to privacy and confidentiality (Ch. 14).

Record keeping

Apart from being important medically, it is very helpful to have a purpose-designed card in the patient's record. This allows all information relevant to contraceptive use to be collated in an easily retrievable form and referred to quickly during any consultation. Many doctors design such cards for their own use and a variety are available from pharmaceutical companies. However the information is recorded it is important that it be regularly up-dated.

Women who have been sterilised or whose contraceptive needs are being adequately met without medical supervision should have a clear entry made on a summary card such as:

1974	Husband had a vasectomy: sperm count negative — date.
or 1978	Prefers sheaths. Used successfully for many years.
or 1979	Hysterectomy.

This acts as a reminder to anyone whom she consults.

Integrated screening

Contraceptive consultations provide an ideal opportunity for health education and screening (Ch. 15). However, it must be made clear to patients that screening procedures, however desirable in their own right, are not directly related to contraceptive use and that contraceptive provision is not dependent on screening being carried out. The majority of patients will view screening as evidence of good care and it will enhance their confidence and satisfaction. However, although they should be encouraged to accept screening, their right to refuse to do so must be recognised and graciously accepted when exercised.

Improving the service

Patients seem to rate care at a family planning clinic highly because they consider it to be more expert and because they have more frequent examinations and cervical smears. They also appreciate more time for discussion, the greater opportunity to see a woman doctor and the convenience of evening clinics at which no appointments may be required.

General practitioner care, on the other hand, seems to be rated highly because of the relaxed and informal service provided, the wide range of consultation times, seeing the same known doctor on every occasion and feeling that the consultation is more confidential. Yet young girls, in particular, often prefer to attend a clinic in another area for the confidentiality that comes from anonymity. In some areas youth advisory sessions cater specifically for this group.

Isobel Allen in her survey *Family Planning, Sterilisation and Abortion Services* (Allen 1981) indicated three areas where patients find general practitioner family planning services to be less than satisfactory. Doctors who provide the service alongside general medical services are sometimes perceived as being too busy to spend adequate time on the subject. On the other hand, where the doctor runs an appointment system, there are consistent complaints about delays in obtaining an appointment. Complaints are also frequently made about obtrusive and aggressive questioning by receptionists and other members of the staff.

While both services fulfil different needs, and between them allow patients a reasonable choice, each could improve some aspects of its function to enhance patient satisfaction.

For general practice this could involve speeding up and facilitating access to the doctor, providing a more comprehensive service, running integrated screening programmes and possibly setting up an intrapractice family planning clinic. Such a clinic would not be a substitute for the present service but an addition to it, to extend its range and the likely number of satisfied customers. A regular evening session staffed by the family planning trained practice personnel, or even a family planning doctor paid on a sessional basis from practice funds, could well pay dividends.

TRAINING IN GENERAL PRACTICE

The Royal College of General Practitioners in its report on family planning (RCGP 1981) states that many aspects of family planning training for general practitioners are unsatisfactory. In particular it indicts inadequate funding of in-service training and the long delay that exists in many places before practical training, particularly in IUD insertion, can be completed.

The RCGP recommends that all vocational trainees in general practice should be encouraged to gain the JCC certificate during their training. It also recommends that practical training should be held partly in the general practice setting under the guidance of general practitioner teachers holding family planning instructor certificates, and not solely in clinics or hospitals as at present. As a large majority of vocational trainees will spend their working lives in general practice, and not in the hospital or clinic situation, this seems very reasonable. It does, however, depend on there being sufficient interested practices and teachers willing to take up the challenge.

CONSULTING SKILFULLY

Although many contraceptive consultations do not make excessive demands on doctors or patients, a significant number do require a considerable degree of sensitivity and skill on the doctor's part. This is particularly true when the patient has attended with an apparently simple physical problem and brings up the question of contraception only towards the end of the consultation.

Consulting skilfully need be neither time-consuming nor laborious and is as relevant to a series of brief interviews as it is to a long initial session. A consultation should reveal not only the nature of the problem or request, but also the patient's attitude to it which in turn will determine whether the doctor's advice is likely to be accepted or rejected.

Ideally, at the end of any consultation or sequence of consultations the patient should feel that the doctor knows everything relevant to her family planning needs. The doctor must:

1. *Be caring and interested.* This is a prerequisite for any successful doctor–patient relationship. If the doctor is thought of as remote or uninterested then the degree of help that can be provided will be diminished. He or she must listen with evident interest to what the patient has to say, rather than ask a series of routine questions which inhibit communication. The patient should get the feeling that the doctor has all the time necessary to concentrate exclusively on his or her contraceptive needs or problems.

2. *Be sensitive to the patient's needs.* The doctor will frequently differ from the patient in age, sex, intelligence, social class and ethnic group, and must make a conscious effort to understand and appreciate the problems, aims and aspirations of the patient.

3. *Allow time when it is needed.* A straightforward physical complaint is easily dealt with by a selective question and answer technique. The patient with contraceptive, sexual or emotional problems, however, may find it more difficult to talk about feelings, anxieties and social strains. Patients themselves may sometimes not realise the true nature of their distress. Strategies have to be developed to obtain the

necessary information while allowing the patient both the opportunity and the time to express feelings and anxieties. An experienced and sensitive doctor may be able to deduce the nature of the problem very quickly and focus on it in a fairly direct way. On other occasions it is better to allow the patient to set the pace and direction of the consultation.

4. *Try not to let one's own beliefs or values interfere with understanding patients with different value systems.* The various cultural taboos which discourage people from talking about many aspects of sexual life present another barrier to communication at the family planning consultation. Doctors are subject to the same emotional constraints as patients and will often collude with a patient to avoid discussing an embarrassing matter. Listening to tape recordings of consultations can reveal how doctors deliberately shy away and change the subject when a patient tries to talk about a taboo area.

Patients often have a great need to discuss many facets of human behaviour and experience, whether or not the topic is likely to embarrass the doctor. Doctors must be honest enough to attempt to analyse their own behaviour to ensure that they are not guilty of being evasive when faced with a subject they find threatening. Very often they rationalise this evasion by convincing themselves that it really was the patient who did not want to talk about it.

5. *Communicate effectively* Most doctors feel that they are good at communicating with their patients, yet various studies (Nuffield Provincial Hospitals Trust) show that between 10 and 90% of patients (average 50%) do not take their prescribed treatment or forget or reject their doctor's advice about changing their habits. Most doctors will benefit from consciously trying to improve their ability to get information over to their patients. To do this the doctor should:

    a. Provide better verbal communication, avoiding medical jargon and pitched at the level appropriate to the individual's understanding and circumstances.

    b. Give out written instructions and information which the patient and the spouse may read and act on in their own time.

    c. Check the patient's recollection and understanding of important points, e.g. 'Now tell me how long should you leave your cap in place after making love'.

    d. Use tape recordings or, even better, video recordings of consultations for self-audit. Critical appraisal of performance will help improve communication skills.

Where patients are receiving all or part of their contraceptive services from community clinics or from colleagues in hospital, good communications either by letter or telephone are of the greatest value in ensuring that they receive the highest standard of care.

SUMMARY

There are many special groups in society who may either have problems in obtaining good contraceptive advice or have difficulties in applying it. The good general practitioner should be aware not only of the patient's organic problems but also be sensitive to cultural, religious and communication difficulties between them.

An open mind, coupled with genuine interest and empathy allows the doctor to be an effective adviser to most patients whatever their individual circumstances and needs.

More women attend their general practitioner for contraceptive advice than go to clinics or elsewhere. Many of them have worries, especially about the pill, about the effects of contraception on their health and how well they may or may not be monitored. A sympathetic awareness of these feelings coupled with an up-to-date and well-planned approach by every general practitioner would help to increase the efficiency and acceptability of modern contraceptive methods.

The provision of a good family planning service to all patients is well within the general practitioner's capabilities.

REFERENCES

Allen I 1981 Family planning, sterilisation and abortion services. Policy Studies Institute,

Fletcher C (Ed) 1980 Talking with patients: A teaching approach. Nuffield Provisional Hospital Trust.

Royal College of General Practitioners 1981 Family Planning. An exercise in preventive medicine. RCGP. London.

Appendix

Contraceptive method	*Fees payable**
Pill	Ordinary fee £8.55 p.a. Claim on FP 1001†
	*At the end of each year a fresh claim can be made one month before or six months after it is due, provided contraceptive care was continuous. Claims may be backdated.
Condom	Ordinary fee £8.55 p.a. Claim on FP 1001. Only claimable if recommended as part of general contraceptive advice, e.g. as temporary postnatal measure.
Diaphragm	Ordinary fee £8.55 p.a. Claim on FP 1001. Claimable even if the general practitioner does not fit it himself (it may be fitted by the practice nurse).
IUD	IUD fee £28.70. Claim on FP 1002.‡ Full fee paid for first year after insertion or IUD change. Ordinary fee payable in subsequent years.
Sterilisation (male and female)	Ordinary fee £8.55. Claim on FP 1001. Only applicable to the year in which advice, examination and follow-up are provided. When advice is given on vasectomy the partner should sign the form.
Periodic abstinence	No fee payable if this is the only kind of contraceptive advice offered.
Postcoital contraception	Fee can be claimed if the GP so wishes.

*This applies to all forms provided services are continuous.
†In Scotland EC102.
‡In Scotland EC103.

Addresses

Association to Aid the Sexual and Personal Relationships of the
Disabled (SPOD)
286 Camden Rd.
London N7 0BJ

Association of Sexual and Marital Therapists
PO Box 62
Sheffield S10 3TS

British Pregnancy Advisory Service
Austry Manor
Wootton Wawen
Solihull
West Midlands B95 6DA

Brook Advisory Centre (Head Office)
153A East Street
Walworth
London SE17 2SD

Family Planning Association
27–35 Mortimer Street
London W1A 4QW

Family Planning Nurses Forum RCN
Mrs Gillian Rands (Chairman)
North Ridge
Byron Road
Maidstone
Kent

324

Health Education Council
78 New Oxford Street
London WC1A 1AH

Institute of Psychosexual Medicine
Lettsom House
11 Chandos Street
Cavendish Square
London W1

International Planned Parenthood Federation
18–20 Lower Regent Street
London SW1Y 4PW

Irish Family Planning Association
Cathal Brugha Street Clinic
Dublin 1

Joint Committee on Contraception
Royal College of Obstetricians and Gynaecologists
27 Sussex Place
Regent's Park
London NW1 4RG

National Association of Family Planning Doctors
Royal College of Obstetricians and Gynaecologists
27 Sussex Place
Regent's Park
London NW1 4RG

National Association of Family Planning Nurses
Mrs Judith Henley (Chairman)
1 Greenacre Park
Law Fell
Gateshead
Tyne and Wear

National Board for Nursing, Midwifery and Health Visiting
for England
Victory House
170 Tottenham Court Road
London W1P 0HA

National Board for Nursing, Midwifery and Health Visiting
for Northern Ireland
123/137 York Street
Belfast BT15 1JB

National Board for Nursing, Midwifery and Health Visiting
for Scotland
22 Queen Street
Edinburgh EH2 1JX

National Board for Nursing, Midwifery and Health Visiting
for Wales
13th Floor
Pearl Assurance House
Greyfriars Road
Cardiff CF1 3AG

National Marriage Guidance Council (Head Office)
Herbert Gray College
Little Church Street
Rugby
Warwickshire CV21 3A

Scottish Health Education Group
Woodburn House
Canaan Lane
Edinburgh EH10 4SG

Scottish Society of Family Planning Nurses
Mrs Mary E Rankin (Chairman)
18 Dean Terrace
Edinburgh EH4 1NL

Women's National Cancer Campaign
1 South Audley Street
London W1Y 3DQ

Index

327